Home Guide to Food Freezing

(Home Guide to Deep Freezing
revised and up-dated)

Audrey Ellis

Hamlyn
London · New York · Sydney · Toronto

Contents

Acknowledgements

The author gratefully acknowledges the help of:
Research editors Gwynedd Hindmarsh and
    Ann Scutcher
Australian Dried Fruits
Blue Band Bureau
Bosch Limited
Butter Information Council
'Carmel' Agrexco (Agricultural Export Company
    Limited of Israel)
Colman's Mustards
Electrolux Limited
English Electric Company Limited
Frigicold Limited
Ann Smith and Frigidaire Division of General Motors
    Limited
Fruit Producers' Council
Kellogg Company of Great Britain
Lakeland Plastics (Windermere) Limited
Mushroom Growers Association
New Zealand Lamb Information Bureau
Porosan Limited
Tabasco Pepper Sauce (Beecham Foods Limited)
Total Refrigeration Limited
The Tupperware Company

Published by
The Hamlyn Publishing Group Limited
London · New York · Sydney · Toronto
Astronaut House, Feltham, Middlesex, England

© Copyright The Hamlyn Publishing Group Limited 1968

First published 1968
Sixth impression 1972
Revised edition 1974
Second impression 1975

ISBN 0 600 34469 X

Printed in England by Fleetway Printers,
Gravesend, Kent

Introduction

Food which is quickly frozen not only remains preserved; it comes out of the freezer full of flavour, bright in colour, with texture unimpaired by a stay of several months, or in some cases even a year. In fact, many foods can safely be stored in the freezer much longer, but most housewives will want to keep its contents on the move. Experimenting to find out just how long you can keep certain items may be fun, but you will certainly want to eat last June's strawberries before the new season's crop arrives, and so on.

Freezing is undoubtedly the best method of preserving food. It is also the simplest. Quick, easy, clean – and certain! Bacteria which cause food to deteriorate are rendered inactive at 0 deg. F. (–18 deg. C.), and since the criterion of successful freezing is to take the food down to –5 deg. F. (–21 deg. C.) within 24 hours, and keep it there, you know your frozen food is safe. There is another reason why quick freezing is so important. All food contains some water which, when it freezes, expands and ruptures the food cells with the jagged ice crystals that have formed. The degree of damage depends on the water contents of the food, and the speed of freezing. The quicker this takes place, the less damage is done.

To get the best results, you need a freezer built for the purpose. The frozen food storage compartment of your refrigerator is intended to keep bought quick-frozen foods, for a period up to three months. The three-star method of marking shows this. Your freezer is intended to freeze fresh produce and home-cooked food so to speak 'from scratch', and keep it safely for periods up to one year.

To help you to distinguish genuine freezers from appliances which are only suitable for storing ready-frozen food, the new international four-star symbol has been evolved. It should be prominently and permanently displayed on all new home freezers. In addition, it is a British Standard requirement that the weight of food which can be frozen daily should be stamped on the rating plate of all new freezers. All this is more fully explained on page 8.

When buying a secondhand freezer, make sure that you are quite clear about what constitutes a freezer. An ice cream cabinet is *not* a freezer; it is a conservator. This means that it will store ready-frozen food at the appropriate temperature, but it is not designed to freeze down food. Unfrozen food placed in it will freeze down eventually, but the temperature cannot be reduced for this to be done quickly. The result may be loss of flavour and texture, which is not really the conservator's fault because it is being asked to do a job for which it was not built. The cabinet you are offered should only be described as a freezer if the temperature can be reduced to –5 deg. F. (–21 deg. C.) or lower, for freezing down. However, secondhand, reconditioned freezers have proved to be very reliable and do yeoman service.

In Chapter Two I have gone very fully into the question of choosing a freezer in the price range you can afford, and to suit your own special requirements. In most cases, you can confidently expect to save the cost of the freezer within three years of purchase. As it is a very robust piece of equipment, it has a much longer life expectation than that.

Meanwhile, you can entertain so much more easily, vary your daily menus considerably, and plan your cooking sprees to give you more leisure time.

The problems of choice, maintenance, and use to the fullest capacity, are all dealt with here in turn. There's even information on how to insure against loss if the contents of your freezer are spoiled through a mechanical breakdown or power cut; and what to do when the freezer is full of food and you have to move house.

You still feel doubtful whether you should invest in a freezer? Let's deal with the questions you are bound to ask, if you've had no previous experience of buying or using one.

First, the cost. This depends on size and capacity, starting with a model of less than 2 cu. ft. at under £50. A really big freezer costs more than £100 but the price ranges on the larger models vary enormously because every supplier is courting the housewife's custom. It is therefore simply a question of finding out where

3

you can buy a freezer to fit the space available and provide the storage capacity you require at the best possible price. For instance, you may be able to buy a top quality freezer at rock bottom price at one of the freezer food centres such as Bejam, who have many branches already, and are opening more. They know you will not be a regular customer for food unless you have a satisfactory freezer.

No expensive installation work is involved. The freezer, like a refrigerator, plugs into any 13 amp socket of suitable voltage.

How much room will it take up? The answer is, if you are really pushed for space – none. With the advent of the refrigerator/freezer it is possible for both pieces of equipment to take up exactly the same floor space in the kitchen – though make sure that the floor is strong enough to take the weight! On the other hand, with plenty of wall space to fill, you might like the luxury of the side-by-side freezer/refrigerator. Or you can choose between the small one-door upright, the large one-door upright, the small top-of-cabinet, the two-door upright and the ever-familiar shape of the chest freezer.

It can be sited in the kitchen, in a cupboard (if well ventilated), in an unused room, in a garage or outhouse – the choice is endless. I have gone into this more fully in Chapter Two, How to choose and install the equipment.

Now for the really important question. Do you actually need a home freezer? The fact that a wide variety of foods can be stored ready for use, for many months if necessary, without losing flavour or quality, makes it a worthwhile buy on the grounds of convenience alone. But this by itself does not make the freezer essential. The advantages, which only appear after you have had your freezer for some time, are of two kinds.

There is no doubt you can effect economies in money. If you have your own garden produce, it costs nothing to freeze it, rather than give it away as surplus. If you have no garden, there are always certain times of the year when weather and other circumstances combine to put a glut of some food on the market at a remarkably low price. In a week or two, or even a few days,

prices are back to normal. If you can take advantage of market-saturation and buy a year's supply of, say, strawberries, asparagus, or cooking apples at a special low price, you must benefit when these items are out of season or scarce and high-priced.

Buying in bulk is always cheaper, and with a freezer to store part of your purchase, you can do this. There are various firms throughout the country specialising in the supply of frozen food in bulk for freezer-owners. Some will deliver straight to your home in insulated vans; and for substantial orders there are usually no delivery charges. At other freezer food centres you will choose your purchases and to carry them home, you will be supplied with insulated bags which keep food safe for up to 3 hours – in most cases, long enough to get it home and tucked away into the freezer. You may find it worthwhile to buy your meat from a wholesale butcher who will cut up the meat just as you want, rather than buying a complete pack from a freezer centre where there is no choice. Retail butchers do supply meat for the home freezer, but don't always specialise in it. The best bargains, therefore, often come from wholesale butchers with a retail outlet.

Now think about the economy in time. Even if you don't go in for bulk buying, you can replan your kitchen activities to save many hours' hard work each week by using your freezer intelligently. To assemble the ingredients and prepare a meat stew may take as much as half an hour. It must be watched, tasted, the heat adjusted, and the stew removed from the heat at the right time or it will burn. The time expended on preparation and dish-washing, for three times as much stew, is not appreciably greater. You can then serve up one portion and freeze two portions for use at intervals of weeks or months if you wish. You'll find in Chapter Four, a method, described and clearly illustrated by photographs, which shows how you can partially freeze the stew in the casserole in which it was cooked, then cut it into portion-size blocks, and wrap for freezing in this shape, so that the least space is needed in the cabinet. Chapter Seven gives many recipes for made-up dishes you can prepare in quantity. All will freeze and reheat successfully.

You will soon learn to exploit your freezer to deal with your own particular catering problems. Maybe children or adult members of the family often require packed meals to take out with them. Preparing and packing sandwiches neatly so that they carry well is a tiresome, time-wasting process, and all the more so when the packed meal is demanded at a moment's notice. It is so easy to prepare a month's supply of various sandwiches to fill the whole family's needs when you have an hour of leisure.

Is there a baby in the house? With a home freezer you can save quite a few pence a week by freezing your own cooked, strained, or finely sieved baby dinners. No busy mother can afford the time to prepare this kind of food freshly every few days. But those tiny cans or jars cost a great deal more to buy than the price of the food they contain. A couple of hours in the kitchen would enable you to prepare three varieties of meat-and-vegetable dinner, including sieving or liquidising, and packing. Then baby is supplied for a month – no precious vitamins are lost, and the cost is halved.

Entertaining is one of life's greatest pleasures to every housewife who loves and takes a pride in her home. In these days of self-sufficiency, it can't be denied that it is hard work. With experience, the hostess can plan the menu in such a way that all the lengthy preparation has been done and the food frozen in advance. In Chapter Eleven there are specimen menus for both formal and informal freeze-ahead meals.

Unexpected guests who arrive hungry, or who love to 'stay on' if they feel they aren't imposing on you to provide a scratch meal, are no longer a problem.

You can also, if you wish, extend the range of your entertaining. Buffet parties for considerable numbers become a possibility on any day of the week, instead of only on days when you can spend most of the afternoon preparing. For a really big affair, you can divide preparing the food into several sessions, and produce it all ready for the final touches, on the great day.

It is only fair to say that you would have to run down your normal stocks of frozen food to accommodate a large amount of party fare, but used in conjunction with your refrigerator, the freezer should supply most of the food needed for a big buffet. Again, there are recipes later on in this book for convenient numbers of people.

To sum up, adding a freezer to your domestic equipment will allow you to cater more economically and more flexibly; to control the output of your own energy in a way you never could before, because cooking is not necessarily a chore you *must* perform for 3–4 hours each day at just the times when it may be most inconvenient. It can be done at times which suit you better; during a quiet spell in the afternoon; in a big burst of energy you may suddenly experience one day; or just when you were able to acquire a quantity of some special produce at peak condition and lower-than-usual price.

To help you adjust to metrication, in the recipe section all quantities are given in both Imperial and metric measures; all spoon measures are level unless otherwise stated. Metric capacity of freezers is measured in litres – to convert cu. ft. to litres multiply by 28·3.

They say the proof of the pudding is in the eating. It is a fact that freezer owners invariably say that they wish they'd bought one of bigger capacity in the first place, and that they can't imagine how they ever managed without it.

Audrey Ellis

Part One
How to choose and use your freezer
Chapter One
The facts about freezing

You don't need to have a scientific bent to understand the principles of refrigeration, for it is all quite easy to grasp. The following simple description of the mechanics of freezing, will probably help you to choose a suitable site and to appreciate better the necessity for adequate care of the freezer.

How the freezer works

The scientific principles on which your freezer works are those of evaporation and compression. A liquid refrigerant, which boils and vaporises at a very low temperature circulates continuously throughout the system, being passed in turn through a compressor, a condenser and an evaporator. This whole system is hermetically sealed.

The actual design of these three components will vary according to the type of freezer (the upright or the chest cabinet); the way in which they work to maintain the necessary low temperature will however be basically the same.

The refrigerant absorbs heat from the freezer cabinet and the food contained in it; heat-laden vapour is then forced by the compressor into the condenser. At this point in the cycle the temperature of the condenser is higher than that of the room and the heat from the condenser is dissipated into the surrounding atmosphere. It's as simple as that. Now let us see how each stage of the cycle takes place.

Compression

The compressor is a pump operated by an electric motor. The compressor motor sometimes makes a slight noise when it is running. It is generally fitted in a separate compartment at the base of the cabinet. The compressor forces the heat-laden vapour into the condenser, making it warmer and warmer, until when the temperature rises above that of the room, the heat passes from the condenser into the room. Because of this heat dissipation and the pressure in the condenser, the vapour is then converted back into liquid and returns to the evaporator. Running of the compressor is governed by a thermostat which is controlled by the temperature inside the cabinet, or inside the evaporator compartment. When this temperature rises slightly the motor operates to start the cycle of compression, condensation and re-evaporation of the chemical refrigerant. It cuts out again when the temperature is sufficiently reduced. This fine control ensures a steady temperature in the cabinet itself.

EVAPORATOR

CONDENSER

Compressor unit

Evaporator coil

Condensing coil sealed to outer liner

COMPRESSOR UNIT

Condensation

The condenser is a series of pipes rather like a car radiator; in the case of the front opening freezer this is located at the back of the cabinet. The condenser of the chest type freezer is usually fixed to the inside of the exterior 'shell' of the cabinet. This is where the vapour is condensed back into liquid.

Evaporation

The evaporator consists of another series of pipes; in the case of the chest freezer this is attached to the outside wall of the interior 'shell' of the cabinet where of course you cannot see or touch it. Where the front opening freezer is concerned, these pipes themselves form some or all of the storage shelves. As the refrigerant, in its liquid form, passes through the pipes it absorbs heat from the interior of the cabinet, boils, and thus becomes a vapour. During this process of evaporation, the temperature of the stored food is reduced.

Cost of running a freezer

It is not the easiest thing in the world to give an exact and accurate figure when working out the true running cost of a freezer. The main items to take into account are depreciation, consumption of electricity and packaging costs.

Freezers do not depreciate rapidly – and most are comprehensively guaranteed by the manufacturer for a period of one year. After this time, the consumers' rights under the 'Sale of Goods (Implied Terms) Act' still apply. So the manufacturers' responsibility doesn't necessarily stop after one year. The life is at least ten years, so that a freezer bought for £50 will presumably deteriorate by the amount of £5 per year, or one costing £150 by £15 per year.

It is hard to be precise on the cost of electricity in running a freezer but the following estimation will give you a general idea of the costs for maintaining a freezer at −5 deg. F. In any case, the manufacturers will let you know the likely running costs of their freezers, which average out to about 1–2 units of electricity per cubic foot per week, or about 18–20p for a 12 cu. ft. freezer.

Certain factors can cause these figures to vary.
1. The amount the freezer is used. Every time the freezer is opened there is a loss of cold air and inrush of warm air. Electricity must be used to reduce the temperature again.
2. The ambient atmosphere. If the freezer stands in a well ventilated, cool place, heat can easily escape from the exterior of the cabinet and it will take less electricity to keep the interior down to the desired temperature.
3. Weather. During an exceptionally long hot summer, more electricity will be used to keep the freezer temperature down, just as it costs more to use an electric immersion heater for hot water during a long cold winter.
4. Design. The amount of electricity used will also depend on whether you have chosen an upright or chest model. The upright type will be more costly because of its front-opening door which releases a considerable amount of cold air, allowing warm air to take its place. Far less cold air escapes from the chest freezer so less electricity is used to maintain the correct temperature.

Packaging is another item one must allow for when estimating running costs. These can be kept down by studying the prices of various types of packaging and using those materials which prove to be efficient and cheap.

The cheaper forms of packaging – such as polythene bags and aluminium foil – will have to be thrown away after being used once, while jars, tubs and waxed cartons may be used several times. You can buy jars and tubs specially for the purpose, of course, but a marvellous way to build up your collection is to save suitable containers, such as used yogurt and cream tubs, for these containers cost nothing, are ideal for storing small quantities of food and will withstand the low temperature of a freezer.

If possible, buy some rigid polythene containers with airtight seals. They aren't cheap, but they can be used again and again and give a first-class means of storing all kinds of food.

For containers or packets without a good seal a special sealing tape, able to withstand low temperatures, must be used. You will also need labels and a chinagraph pencil or gem marker. More information on packaging is given in

7

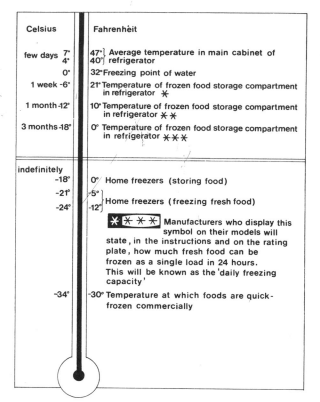

Celsius		Fahrenheit	
few days	7° 4°	47° 40°	Average temperature in main cabinet of refrigerator
	0°	32°	Freezing point of water
1 week	-6°	21°	Temperature of frozen food storage compartment in refrigerator ✳
1 month	-12°	10°	Temperature of frozen food storage compartment in refrigerator ✳✳
3 months	-18°	0°	Temperature of frozen food storage compartment in refrigerator ✳✳✳
indefinitely	-18°	0°	Home freezers (storing food)
	-21°	-5°	Home freezers (freezing fresh food)
	-24°	-12°	
			✳ ✳ ✳ ✳ Manufacturers who display this symbol on their models will state, in the instructions and on the rating plate, how much fresh food can be frozen as a single load in 24 hours. This will be known as the 'daily freezing capacity'
	-34°	-30°	Temperature at which foods are quick-frozen commercially

Chapter Five, beginning on page 63.

Temperatures for food preservation

While most people realise that a cold temperature will preserve food, few of them know the precise temperatures at which food may be safely kept and for how long. A domestic refrigerator, for example, maintains a temperature of some 40 deg. F. in the cabinet. An efficient frozen storage compartment maintains a temperature as low as 10 deg. F. which will keep commercially quick-frozen foods frozen for any time up to 1 month for a two-star refrigerator. A frozen food compartment with its own door, separate from the door of the main cabinet, can maintain 0 deg. F. permitting storage of, perhaps, 50 lb. frozen food up to 3 months.

However, a domestic freezer not only stores food at still lower temperatures (usually between 0 deg. F. and −5 deg. F.), but can be reduced down to −10 deg. F. or less, to ensure quick freezing of foods in bulk from a normal room temperature, to at least as low as 0 deg. F.

Briefly, the domestic refrigerator cabinet is fine for tiding you over from one shopping day to the next, keeping perishable foods safe for up to a week, drinks and salads pleasantly chilled. The freezing compartment will turn water into ice-cubes, and keep them frozen; it will preserve commercial frozen foods from thawing, for a limited time. A separate freezer compartment is more efficient and will extend the storage life of bought frozen foods. To help customers distinguish a freezer from appliances designed only for storing frozen foods, the new four-star symbol has been introduced and will be found prominently displayed on the cabinet of new freezers.

To justify the four-star rating, the manufacturer must state in the instructions and on the rating plate (to be found at the side or back of the cabinet), how much fresh food can be frozen as a single load in twenty-four hours. You should not try to freeze more than the capacity given on the rating plate. This is usually about 10% of the freezer's total capacity although there are some freezers that will freeze down a little more each day.

Unfortunately, here is a case where time is of the essence. Only *quick* freezing preserves the colour, flavour and texture of the food adequately. For this, and to freeze foods in any quantity, you need the extra freezing power of the home freezer, which freezes quickly, and maintains a low enough temperature to keep many foods for months, even up to a year. A commercial freezer drops the temperature even lower than the domestic freezer and can preserve food for years.

Experiments have proved that colder temperatures increase the period of safe storage and that quicker freezing preserves the natural qualities of the food better than slow freezing.

If you study the diagram to the left, you will see that water freezes at the surprisingly high temperature of 32 deg. F. That is why even a low-powered refrigerator can make ice cubes, and why, if you turn the regulator in your own refrigerator cabinet down to minimum, your milk and salad greens form ice crystals. It is the water in these foodstuffs which is freezing, and a much lower temperature would be needed to freeze, for example, a joint of meat.

Obviously, the safe storage period increases as the temperature falls. Inevitably, there is a snag, for it takes more electricity to keep the temperature lower, and therefore puts up the running cost. What then is the maximum temperature at which most foods can be stored successfully for a reasonable period of time? This varies from one food to another; but one thing does not alter – the fact that a cold temperature slows down the chemical action of bacteria and fungi. Taking into account that you may wish to store such different foods as fresh fruits and vegetables, raw meat and poultry, made-up dishes, cakes and bread, the maximum safe storage temperature is 0 deg. F. or less; preferably −5 deg. F. Some of these foods will keep successfully for only a few weeks, others for months, a year, or more. But this is an economic temperature which housewives experienced in using home freezers find suitable for their needs, without the running cost soaring sky high.

When you set out to buy your own freezer, you will find that there is a choice between the upright type and the chest type.

The upright freezer

This type must be used very carefully to limit the inrush of warm, moist, ambient air from the room, each time the door is opened. There is a disadvantage in a freezer with a door which takes the place of one of the long walls, for it exposes the whole of the frozen food to warm air as soon as the door is opened. Alternatively it is possible to buy a two-door upright freezer. As only part of the freezer is exposed when one of the doors is opened it keeps the warm-air intake to the minimum. A loss of cold, dry air, and the inrush of warm, moist air means that extra electricity must be used to restore the cold, dry atmosphere necessary. Cold air is denser and heavier than warm air. When the door is opened the cold, dense air tends to flow out at the bottom and lighter, less dense, warm air to flow in at the top.

Vertical doors must be very well fitting, or there will be a continual slight loss, as well as the noticeable loss when the door is opened. On the other hand, since it is easy to reach any pack, no matter where it is situated, in the upright freezer, the door need not be kept open for long. A moment's thoughtful consultation of your plan of the freezer's contents before you open the door, should enable you to find what you require in a few seconds. This method will save you a considerable amount of electricity over the years.

The chest freezer

Anyone who has fished frantically around in this type of freezer in a shop, searching for an elusive packet of peas or chips, knows its disadvantages. However, the fact that it is left *uncovered* in the shop demonstrates the great advantage as far as conserving cold air is concerned. The chest freezer is, basically, a box with a lid, just as the upright freezer is a cupboard with a door. The domestic chest freezer is more economically run because it has a lid, which can be kept closed except when packs are actually being put in or taken out. Yet there is surprisingly little loss of cold air even with the lid open. When you open the lid, draughts of turbulent air invade the freezer, penetrating a few inches below the top but, unless it is a very strong draught, no further. The cold, dry air towards the bottom of the cabinet is much denser than the warm, moist, outer air of the room. It is therefore only disturbed to a limited extent. If this were not so, the open display cabinets seen in shops would be both inefficient and expensive to run. However, even a light draught of warm air does cause the compressor motor to start operating. Food stored near the top of the freezer will thaw slightly. And the moisture contained in the warm air will be deposited in the form of frosting round the inner surface of the cabinet. For all these reasons, you should not be tempted to leave the lid off longer because you are used to seeing open chest freezers in the shops.

The careful planning of the arrangement of packs in the chest is important, because you will always have to remove the top layers to get at the packs further down in the chest. Baskets or other containers which can be quickly lifted out to allow access to the lower half of the chest are not only a convenience, but save money because you will not have to leave the chest uncovered so long. The baskets must be of a size and shape you can lift without strain when they are fully loaded. These can be very heavy when they are full, so try to arrange that the packages which are required most frequently, are loaded into the top basket so you can reach them easily.

The storage capacity of the freezer

The average amount of food stored in a home freezer is 20 lb. of frozen food to every cubic foot, but this can vary enormously, depending on the way a freezer is packed. If all the food was stored in regular-sized, tightly-packed 'bricks', much more could be squeezed into the freezer. But food isn't stored like that and, in two important ways, it is as well that it isn't. In the first place, air circulation is needed for efficient running of the freezer and in the second, there should be finger spaces between the packages of food, or they would tend to freeze together into one solid block, making removal difficult, if not impossible!

A freezer can only freeze a limited amount of food at one time. You will find that all freezers with the four-star symbol, will have the weight of food that can be frozen daily stamped on their rating plate. Remember that attempting to freeze too much fresh food in one day may unduly raise the temperature of the food already in store. Positioning is also of the greatest importance. Packs to be frozen should be placed against a refrigerated surface, usually walls or floor of the cabinet, but in the case of some upright models, the shelves themselves are part of the refrigerating system. You can get a more precise idea of the freezing rate, if you allow from 2–6 lb. of food per cubic foot of storage capacity to be frozen every 12 hours. This variation exists because of the different constituents of one food or another. Most fluid foods take longer to freeze and so cannot be loaded as fast as dry foods, such as poultry and game. A rate of 2 lb. per cubic foot,

therefore, refers to the fluid foods, while 6 lb. per cubic foot is the rate for foods such as game. The majority of foods such as meat, fruit and vegetables, are neither too dry nor too fluid, so these are given a rate of 4 lb. per cubic foot.

The following table will give you some idea of the storage capacity of your freezer, but the guide cannot be completely accurate, not only because of the different ways a freezer may be packed, but because the size of the freezer baskets and the arrangement of the shelves make an appreciable difference too.

Into each cubic foot of space you can get:
16–20 1-pint tub-shaped cartons
20 lb. meat or poultry
25–40 1-pint square or oblong cartons

Dealing with emergencies

Luckily, there are few occasions when the electricity power fails, but even so it is wise to have ready an emergency plan.

Faults in the local electricity supply are usually dealt with quickly and power restored, at worst, within a couple of hours. Failure in the working of the freezer may take longer to deal with even though most refrigeration engineers run a 24-hour emergency service.

The thought of coping with £100-worth of mouldering food is bad enough, but think how much worse it would be if you also had to stand the financial loss. The answer is to insure the contents of your freezer. An insurance broker can arrange for this to be added to your existing household policy, or can organise a quite separate policy. Insurance should cost about £3 a year to cover both the freezer and £100-worth of food against power failure (though not usually strikes), breakdown or theft, but does not cover negligence, like forgetting to switch on the power. Remember the cautionary tale of a friend of mine, who switched off the freezer so that she could plug in the record player during a barbecue party and only remembered the freezer a week later! Some companies operate a 'block' scheme to provide insurance for a group of stores or retailers. So your supplier might well offer you a package deal of freezer and insurance combined.

Once the power fails, the food in the freezer will begin to thaw, but how long this will take depends on a number of factors. Thawing is quicker and more complete during a heatwave when the air from a room at 75 deg. F. or more, gradually raises the inside temperature of the freezer. During a cold spell, the ambient temperature may be very much less than this so that once the freezer warms up to about 15 deg. F., the rate of thawing will greatly reduce. A freezer packed full of food will take longer to thaw than a half full one, and in a chest freezer, the packages at the bottom will probably take twice as long to thaw as the ones near the top.

Whatever the circumstances, you can reckon on having a few hours grace to deal with the emergency before the food begins to thaw – say up to 12 hours during hot weather, and in cold weather you may have as much as three days. This means, of course, that no action need be

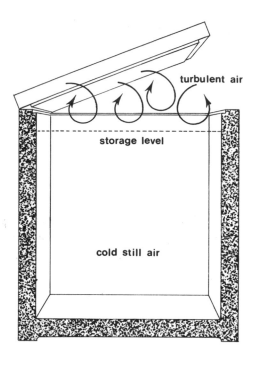

turbulent air

storage level

cold still air

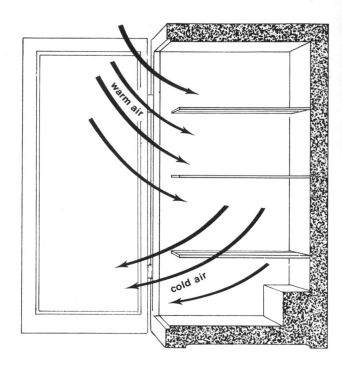

warm air

cold air

taken at all when the electricity is to be restored within an hour or two, as is usually the case. But if possible don't open the lid or door, or warm air flowing in will speed thawing.

The use of dry ice during a power failure, or emergency which necessitates switching off the freezer, is not recommended. Dry ice causes such a drastic reduction in temperature that it can fall low enough to damage the mechanism and the interior of the cabinet.

The happiest solution for saving food would be to transfer it to another freezer – and this isn't such an impractical suggestion as it may seem. A friend or neighbour, who hasn't had an electricity cut, may have some spare storage space in her freezer, or the local butcher may be willing to store your food until your freezer is again working normally.

If you are a regular customer of a frozen food supplier, enquire whether he would be willing to help during an emergency by temporarily storing the entire contents of your freezer. In any event, the important thing is to foresee any emergencies and form a definite plan to deal with them before they occur.

If you set your domestic refrigerator to its very coldest it should be just cold enough to maintain food that is already frozen until the emergency is over, though it will not be cold enough to actually freeze the food. In any case, decide on your emergency plan in advance.

Preparing a new freezer for use

After installation and testing, switch off and wash inside with plain warm water, dry thoroughly. Don't adjust the thermostat as it will be pre-set to the required temperature unless it has variable settings to be set after cleaning. Switch on, leaving for 12 hours to chill the cabinet thoroughly before use.

Moving house

When you move house, you may well be faced with the problem of how to transport a well-stocked freezer. If you move to a distance, it is safer to let the contents of the freezer run down. Check that the removers will undertake to handle the cabinet loaded; it is, after all, a very weighty item. If it must be unloaded, contents can be packed in tea chests with dry ice, if available. If the move will be completed within a day, you may be able to manage without using dry ice. Make sure the freezer is the last item to be put on the van, and will therefore be first off at the other end. Check that the plug will fit the new socket so that you can plug in and switch on the power at once. If your removers insist that the cabinet must be emptied, line a packing chest thickly with newspapers and a blanket (first chilled in the freezer before switching off) and pack the food tightly in layers. It is worthwhile remembering that the larger the number of frozen packs stacked together, the longer they will stay frozen and deterioration of food should not occur within a 6 hour period. However, the packages which were stacked at the top of the chest should be used within one week.

A few words of warning. The ideal freezer installation is that which is on a completely separate circuit. This allows the remainder of the current to be switched off, when the householder is away. The cabinet should be connected to a suitably earthed outlet (usually 13 amp) and protected in all cases by the correct fuse. The freezer motor can cause a surge of power when switched on which could possibly blow a fuse if plugged into a 5 amp socket. It is therefore wiser to connect all freezers – even the smallest ones – to a 13 amp outlet.

Preventing crystal formation

In every stage of deep freezing, speed is of the utmost importance. Once picked or bought, the food destined for the freezer must reach the kitchen quickly, be prepared without delay and put into the freezer as soon as it is ready – all of which helps to slow the deterioration that begins as soon as a fruit or vegetable is picked, or an animal killed, and so on.

Speed in the actual freezing is also essential but for a slightly different reason. All food contains some water which, when it freezes, expands and forms jagged ice crystals; these rupture the food cells. When freezing takes place slowly, more ice crystals can form, so greatly increasing the amount of damage to the food cells – particularly marked when the water content is high. This is why the temperature

when freezing food should be as low as −5 deg. F., and lower still if possible. An ordinary refrigerator, even one that seems exceptionally efficient, just doesn't produce low enough temperatures in the freezer compartment to freeze food quickly, although it can maintain food in its frozen state for up to 3 months. For example, if you are de-frosting the freezer, you can safely leave some frozen food in the coldest part of your refrigerator for several hours.

Another way to ensure that the food is frozen is to pack it in fairly small parcels. A whole lamb would take much longer to freeze than if it were portioned.

A home freezer is only able to freeze a limited number of packages. Don't put in more during any one day than recommended on the rating plate of your freezer, for this also slows down the process. Simply wait until the packages are frozen, then transfer them to the main storage areas of the freezer, leaving space for a further batch of food to be frozen.

The 1·75 cu. ft. Hoover Home Freezer can be reduced to −18 deg. F., to ensure really quick freezing, and freezes down 16·5 lb. of food at one time.

Chapter Two

How to choose and install the equipment

The choice of a site for the freezer is extremely important. It should be dry, cool, and reasonably well ventilated. It has to be remembered that the freezer actually warms the surrounding atmosphere, and thus the room in which it is placed.

Where to place the freezer

Let's suppose that you have the choice of putting it in a small larder, where it will be a tight fit, or outside the house in a garage. It will be less conveniently to hand in the garage, but it will cost less to run, and probably function more efficiently, providing the garage is dry and the temperature doesn't drop below 40 deg. F., even during the coldest weather. This may sound complicated, but the reason is quite easy to grasp.

At temperatures below 40 deg. F. the motor cuts out for long periods and, because the action of the motor is spasmodic, the temperature of the surrounding air is never raised. This causes a chain of reactions: the motor cuts out for longer and longer periods, allowing the temperature inside the cabinet to fluctuate considerably; when the motor is operating, the energy which it creates meets the very cold atmosphere, causing condensation and subsequent damage to the motor and casing of the freezer.

On the other hand, heat from the surrounding atmosphere (in the room where · the freezer stands) penetrates the insulating walls of the freezer. It also invades the cabinet every time the door is opened. The temperature within the cabinet rises, and the thermostat, which controls that temperature, brings the motor into action. The heat which has penetrated the freezer's 'defences' is repelled and 'pumped back' into the room. This in itself does not raise the temperature of the room, but the electrical energy used in the 'pumping' process does. This energy is dis-charged into the room in the form of heat, via the compression unit which, as you can see in the chart on page 6, is situated outside the storage area of the cabinet.

In ideal circumstances, in a dry but well ventilated pantry, the ambient temperature will be only high enough to cause the refrigerator unit to operate briefly, from time to time, keeping down the temperature level inside the cabinet to the degree set on the thermostat. But suppose the freezer is sited in a small larder, in summer, where the temperature, at the hottest time of the day, may rise to the eighties. And suppose you frequently open and close the lid or door, thus allowing an inrush of very hot moist air? The refrigerating unit may be operating almost all the time, to keep the interior of the cabinet as cold as the thermostat decrees it should be. The amount of electrical energy used and the heat discharged may be equal to that of two or even three powerful electric light bulbs, or almost as much as a very small electric fire.

Here, ventilation is all important. If the room is well ventilated the heat is soon dispelled, but if the freezer is sited in a confined space you will create an undesirable pattern – heat discharged which cannot be dispelled, thus causing the refrigerating mechanism to function more often and discharge more heat. In fact, the problem may be so acute that in hot weather the refrigeration unit is working continuously, and even then the room temperature does not remain constant, but continues rising. Eventually, a point will be reached when the temperature in the room is constant, but the temperature inside the cabinet is not as low as it should be according to the thermostat setting, because the refrigeration unit is working to capacity, and then the frozen food will start to thaw slightly, with the attendant risk of spoilage.

To gauge whether the site you have chosen is suitable, imagine the result if the room were constantly lit by three 100-watt bulbs. Would the room temperature rise appreciably? If the space is so confined that the temperature would be noticeably raised, it will not be the most economical place for the freezer. If, however, a small larder or even a cupboard is the only possible site for your freezer (which in a modern house or flat may be the case) you need not worry that the installation of a freezer would be *unsafe*. It will merely be a little more expensive to run and deteriorate more quickly.

Dampness may cause some damage to the motor, and by condensing on the outside of the cabinet, eventually damage the exterior. The condenser cannot do its job of removing heat efficiently unless there is adequate air circulation round the cabinet.

Do consult a refrigeration engineer about the best place in your home for the freezer. He will probably tell you that a cellar, garage, out-house or box-room will be better than your warm kitchen, because they are *colder* and therefore more suitable, provided they are *dry* and *well-ventilated*. (However, remember that it should not be put in an out-building where the temperature is likely to drop below 40 deg. F.) A model fitted with a lock is desirable in an out-house or garage. If a freezer is placed in an out-of-the-way spot and is at any time inoperative, take the precaution of keeping it tightly locked. Or, if it is out of action for a really long time, remove the rubber gasket that ensures an air-tight closure to the lid or door. It has been known for young children to play inside an unused freezer and should one child lock another one into the freezer, a tragedy could occur.

If the kitchen proves to be the most suitable site for a freezer, the refrigeration engineer will recommend an area of the kitchen that is well-ventilated, even if it is warmer, rather than one which is larger and colder, with poor ventilation. He will also advise on the suitability of the socket, plug etc., and the correct voltage. This is no more of a problem, however, than plugging in an electric cooker or refrigerator, and has already been discussed in Chapter One.

How heavy is a freezer?

The question of weight must be carefully considered, especially in an old house where there may be a cellar underneath the kitchen floor. This applies especially to upright models where the tendency is for a stacked freezer and refrigerator. The loaded weight of a refrigerator on top or underneath a freezer may place considerable strain on a small floor area. For example, an area about 23 × 25 inches may have to bear a total weight (when both refrigerator and freezer are fully packed) of over 500 lb. Chest freezers have a much wider base. For example, a chest of 6·2 cu. ft. capacity, with a floor area of 30 × 25 inches weighs 150 lb. unladen and 270 lb. laden. A 9·5 cu. ft. capacity chest with a floor area of 40 × 25 inches weighs 182 lb. unladen and 372 lb. laden. Where the concentration of weight falls on a very small floor area, the construction of the floor itself must be adequate to take this strain. Professional advice on this point may be needed.

A freezer in the kitchen

Some typical kitchen layouts showing where and how a freezer could be fitted into the space available, in the handiest and most economical position, are shown overleaf.

The type of freezer you need

The floor space available may limit your choice to an upright cabinet, or a small chest cabinet. However, if you have a slightly larger area of floor space available, you should consider carefully how the freezer will serve your needs and how convenient it will be for you personally to use.

For instance, if you are short, you may find it difficult to reach right down into a chest cabinet. An answer might be to look out for a chest that is a little lower than usual – for example, the Bosch which is 33½ inches high, 2½ inches shorter than the average freezer. On the other hand, if you find it hard to handle heavy weights, remember that you may have to remove baskets filled with packs of frozen food to get at items near the bottom of the freezer. You will not be able to lift this weight as doctors tell us we should, that is, by bending your knees and keeping your back straight as you

lift the basket. In fact, you will have to bend forward and take the strain of the weight with the maximum of muscular effort instead of the minimum. If you intend making very frequent visits to the freezer, constantly repositioning the contents, bear this in mind. However, if you open and close the freezer often, less cold air is lost from the chest type, making it slightly more economical to run.

Upright freezers

When space is limited, the upright cabinets come into their own. They can be tiny (there is one of less than 2 cu. ft.) or they can reach the ceiling – still taking up no more floor space than the average refrigerator. Indeed, if the freezer is sited underneath an existing refrigerator, it will take up no extra space at all. If yours is a modern fridge, there may be a stacking kit available.

Alternatively, you could have a look at a two-door refrigerator/freezer with equal space in each storage compartment, or one with less refrigerator space. It is becoming obvious that the space ratio of a 4 cu. ft. refrigerator over a 10 cu. ft. freezer suits many people's needs.

Other upright models include a small one-door freezer which provides its own working surface, or fits under it. But try to be sure that the model you choose has one removable shelf or you will have trouble storing away bulky packages.

The large one-door upright has plenty of extra space but the loss of cold air here is considerable. Better, perhaps, is the large two-door model which, because only one door opens at a time, conserves cold air more efficiently. Some models have shelf doors which help keep the cold air where it should be – inside the freezer! Other two-door freezers have the two compartments completely separated.

The door space of the upright models is particularly useful, especially for tiny packages which tend to 'disappear' in the main body of the freezer. But with the loss of cold air when the door is opened, some of these door-stored items can thaw slightly so that you must be careful what food you keep there.

Real luxury is the side-by-side refrigerator and freezer unit – ideal for those lucky enough to have plenty of wall and floor space to spare in the kitchen. Or you can have a matching 'companion' freezer and refrigerator which stand side by side.

The baby freezers of 1·75 cu. ft. capacity continue to be very popular with people having minimum space. They freeze down very efficiently and, although they are rather too small for the average user, extra space can be saved by transferring frozen-down food to the frozen food compartment of a three-star fridge where it can be kept for up to three months.

Chest freezers

The chest freezer has marvellous storage capacity and is more economical to run than the upright because of its top opening lid; its only real disadvantages are the difficulty of reaching the bottom, and the extra space it takes up.

I am afraid you can do nothing about the floor space it needs; but you can cope with the problem of depth. A system of sliding baskets can be suspended from the rim offering layered storage. Clearly labelled or colour-coded batching bags make for easy identification, so that you are not scrambling about looking for an elusive pack. And if you can keep a log which, among other things, shows where you can find a particular item, this will also help limit the amount of time you are bent double over the freezer.

It simply remains for you to decide exactly how you want to use your freezer. If you wanted it only as an extension of the frozen food compartment of your refrigerator, a smaller one may be quite adequate for your needs. But remember that most people wish that they had bought a bigger freezer in the first place. And for the needs of the average-sized family a freezer of at least 10 cu. ft. is deemed essential, or 2 cu. ft. per person plus 2 extra cubic feet for occasional entertaining.

After all, to prove an economical venture, it should be possible to take advantage of bulk buying and freezing down one's own food. And without a big enough freezer, this will not be possible.

The economics of home freezing

Most of us have to consider the purchase of a large piece of domestic equipment from the economic point of view. How much money will have to be invested? How much time and effort will the investment save? Perhaps most important of all, will the investment be in itself a money-saver, and thus in time pay for itself?

If sheer value for money is your criterion it is always more advantageous to buy a larger freezer than a smaller one. While it is true that the cost of the freezer rises roughly in proportion to its storage capacity, a small freezer definitely costs more per cubic foot than a large one. The large one is therefore a better buy in terms of space for money. The same rule applies to running costs. Again, the consumption of electricity rises in proportion of the size of the freezer, but small freezers consume slightly more electricity per cubic foot, than large ones.

Some pieces of equipment that cost around the same price as a freezer only save time and effort, which has considerable value. The refrigerator may claim to save you money in spoiled food and thrown-away left overs. But there is a really dramatic money-saving factor in the ownership of a freezer. Here are three typical stories from housewives in widely differing circumstances, all of whom have made a different choice and use of their freezer who yet agree that they are making a very substantial saving. They all find a great advantage in cooking when it happens to suit them rather than just before a meal, and that they enjoy more varied menus than in pre-freezer days.

CASE HISTORY no. 1

"Basically, I am away from home for almost twelve hours a day, five days a week, as I am the editor of a women's magazine. Even so, since I live in a village where there is a busy round of social life, I like to take my turn in giving dinner parties, buffet suppers, charity coffee mornings, tea parties for the old people and so on. These may well – and usually do – come at a time when I am most heavily engaged at the office – on press day or, even worse, when an unforeseen crisis has cropped up.

On the other hand, because of the rush and tear nature of my work, I occasionally like to spend a whole Saturday relaxing in the best way I know – kneading, rolling, whisking, stirring, baking or at least preparing batches of cakes, pies, flans, bread, purées, pie fillings, pâté, etc. One of the advantages of living in the country is having a big garden and, naturally, the opportunity of freezing one's own fruit and vegetables, i.e. raspberries, strawberries, etc.

Shopping, too, is a problem with the unpredictable working hours I have, and a visit to the Soho street market and delicatessen shops is a luxury I can enjoy only about once or twice every two or three months.

For all these reasons, a home freezer has now become a necessity and in no way a luxury in my household. In fact, it saves me money.

Lots of people buy a small freezer only to find that they need one much larger, but I was lucky enough to discover this before committing myself to a small one. A friend was spending two years abroad and offered me the use of her freezer until she returned – I leapt at the chance. It was a 6 cu. ft., the size I would probably have chosen to buy for myself. I knew very quickly that I needed a freezer twice as big, for once I had packed in fruit and vegetables from the garden, I hardly had room to squeeze in a pound of sausages! In fact I bought a much larger chest freezer of 15·5 cu. ft. capacity which I considered to be an economy.

When I go to Soho, I buy two or three pounds of red and green peppers when cheap, at least a couple of pounds of button mushrooms and as many aubergines as I can carry. Then, that evening, I will wash, de-seed and cut up the peppers and put them into plastic bags; wash, dry and pack the mushrooms, and sauté the aubergines in butter and store them in plastic boxes. There I have the ingredients for casserole flavourings, moussaka, omelette fillings and soufflé mixtures all ready at hand. Our own garden provides plenty of onions and shallots and these I mince and store in the freezer in large quantities at a time. A spoonful flavours a soup or stew, a cupful makes onion sauce as an accompaniment for lamb, and all in moment.

Similarly, I make two or three large bowlfuls of

apple purée each year from the garden windfalls and use it for pie fillings, sauce, soufflé and many other dishes.

The largest pumpkin we grow each year is always destined for the freezer. I make a traditional pumpkin filling complete with spices and beaten eggs – right up to the stage when it is ready to fill the flan case – and store it in plastic boxes with snap-on seals. Six months after the last pumpkin has been picked, I get a special thrill at producing a pie made possible by last autumn's golden harvest!

I have to be in the mood to make pâté – I go through spells when the sight of all that liver makes me feel positively ill – so when I am, I make five or six complete terrines, decorated with streaky bacon, and a spray of bay leaves from outside the kitchen window. Once defrozen, these provide an instant first course for a dinner party or have even taken the centre of the table for an impromptu buffet party.

One of our favourite dishes is curry, authentically prepared with spices which we grind ourselves. This, again, takes a great deal of time and trouble and so I make at least three times as much as we need at once. Curried meat or fish, vegetable dishes and dhal are stored in separate boxes ready to join the traditional quickly prepared accompaniments.

Actually, I firmly believe that the curry improves immensely by being frozen – and I would choose to give a de-frozen curry to anyone I wanted to impress.

Soups take up a lot of space in the freezer, but I always have some as a stand-by (my pride won't let me buy a can or packet!). We grow artichokes by the bushel, and find that few of our friends meet this soup elsewhere. Every winter, therefore, I freeze four or five pints of artichoke soup, made with the richest chicken or turkey stock I can produce, and store it for use in the summer. Served chilled, with a stir of fresh cream, it is a talking point for ages.

In the summer I always buy a largish quantity of fresh salmon 'pieces' from a London department store. This I make into mousse, in deep flan rings which I remove to put back into circulation after freezing. A last minute decoration with cucumber or asparagus is all that is needed for another party dish.

As my husband shoots, I always have to leave room in the freezer for anything he may bear triumphantly into the kitchen. The less attractive joints of rabbit or hare I plain-boil for the cats and keep as an insurance against ever forgetting to buy their rations. Fricassée of rabbit, jugged hare, casseroled pheasant or partridge (cooked to a coq-au-vin-type recipe) then go into the cabinet for an emergency call.

Since we live a tolerable journey away from an East Coast fishing port, we sometimes drive there and stock up with fish bought wholesale. First, I boil several pounds of the cheapest varieties for the cats, then I pack others raw, in foil, and plastic bags. White fish frozen in a cream sauce is a good reserve, but does, I find, lose some of its subtle flavour."

CASE HISTORY no. 2

"It's not that David is an unreasonable man, after all he was fairly calm when I took off the left gate post with the back bumper. But mention the word 'freezer' and he would go spare. I would present him with all kinds of carefully worked out facts which proved without question that he would save absolutely pounds. And my husband would have ready all the facts to prove that I was completely out of my mind.

I was astonished when he finally gave way because I still don't think he was convinced over the value of having a freezer. Perhaps he was after a quiet life!

Anyway, we marched into our local freezer centre one Saturday afternoon and came away an hour later, the proud owners of a double-door upright 13 cu. ft. freezer.

My chief aim was to prove to David that his fears were unfounded and that I could run the freezer economically. I investigated all the various sources of bulk buying – commercially-frozen foods, meat and fresh fruit and vegetable produce I could freeze down myself.

We live in the country, so it is quite easy to buy pick-them-yourself strawberries and raspberries in season, while there is another farmer nearby

19

growing the most lovely runner beans and he lets us have them at a very good price.

Our freezer centre has a huge range of goodies to tempt me, but I have to keep a tight rein on myself before I succumb to a spending spree that would make my husband feel very smug! Instead, I have compared the prices there with those at other local centres and have found that it is possible to make good savings by shopping around.

I was always a bit scornful of people who trudge from supermarket to supermarket, picking up the odd bargain – saving 10p on each shopping trip and using 11p-worth of shoe leather! But when the savings made on a giant pack of fish fingers (my son's staple diet) could be as much as 50p it didn't seem so silly after all. But at the same time there wasn't much point in buying a big pack of beefburgers – even though I could save quite a bit – because only David likes them and it seemed very wasteful of freezer space to store them when the space could usefully be occupied by something else. I resisted the temptation to stock up with monster packs of vegetables when it would be cheaper still to wait till the new crop came in and freeze down my own. The children and I found a new interest in gardening, and began to grow all kinds of basic vegetables for the freezer. I found certain varieties mentioned in the seed catalogue as 'good for freezing' and made a point of buying them, which made me feel very professional about the whole thing. This kept us going for much of the year, and only when our home-grown vegetables were used up did we start buying them from the freezer centre.

I made the mistake of the year when I bought my first meat. I ordered a half-pig from a local butcher and I assumed it would come ready-portioned. Now I never assume anything! It arrived at the back door, shrouded in muslin, and was left on the table while I stood there gaping. I lifted a corner of the muslin and nearly died when I saw the pig returning my gaze unflinchingly! It seems I was the lucky lady to win the half with the head.

The charge the butcher would make to have the pig back for portioning was phenomenal, so

David and I battled with it all one evening. Never again! After that I found a wholesale butcher who could offer excellent prices on ready-portioned meat delivered free of charge, who cuts up the joints just as I want them.

It goes without saying that David is even more of a freezer addict than I am these days. But if you take a peek in my freezer, I must admit that you will see the odd pack of Coq-au-vin or Crab Newburg nestling in the depths. I still follow my Freezer for Economy edict, but David is the most dreadfully compulsive buyer when we visit the freezer centre . . ."

CASE HISTORY no. 3

"I have now been a working housewife for several years. When I first decided to buy a freezer, I was spending three hours a day, in addition to my working hours, travelling to and from London.

Inevitably, our evening meal consisted of some type of chop, usually frozen, frozen vegetables and vast quantities of potatoes to fill up the corners.

As a professional cook I found this way of refuelling uncreative, monotonous and very expensive.

Having decided that something must be done, I started looking at literature on different freezers.

My decision to plump for a small upright freezer was based on two factors. The first consideration was space. In a 'business couple's kitchen' there is barely space for a cooker, let alone a freezer. The only place possible, was the closet leading off the bedroom. Curiously enough, this has proved to be the ideal site.

The closet is situated under the eaves of the roof, three sides having outside walls, the fourth being the entrance from the bedroom. The closet is cool without being too cold, keeping running costs down to a minimum.

Consideration number two was that, in my busy life, I could not afford to have machinery which required a great deal of labour and time. I have always insisted that all equipment in my kitchen must work for me – not vice versa.

Then came the task of stocking up; this was planned like military manoeuvres. I decided that the freezer must carry sufficient food of different varieties, so that I need cook only one batch of food, enough for ten servings, each weekend. By inflicting this weekly chore upon myself, we were able to eat freezer-meals from Monday to Friday, without depleting the stocks.

Due to my having to leave home shortly after eight each morning, and not returning until after seven, I found shopping almost impossible. The most difficult thing to buy was bread. Invariably, it was all sold by lunch time. A monthly trip to the family baker solved that problem. I decided against bulk buying frozen foods – I'd had enough chops and fish fingers. Instead, I made a batch of pizza and cooked them in Swiss roll tins, before cutting into squares, for easy packing.

Sauces were the next priority – one quart of tomato sauce for meat dishes; a quart of Bolognaise, cooked for many hours, for quick meals and a quart of well flavoured, white sauce. This is always useful as a base for cheese, egg or parsley sauce and can also be used for vol-au-vent fillings. I flavour with lemon juice, anchovy essence and tomato ketchup for a fish filling, or use Worcestershire sauce, mushroom ketchup and cream for a filling of chicken and ham.

I always have an abundance of casserole-type dishes, such as goulash, steak and kidney, curried chicken and rabbit with vegetables. These dishes provide a certain amount of variety; they are far more economical than speedily cooked chops and steaks; they are easy to reheat and can be 'dolled up' sufficiently to impress guests, provided one has a bottle of wine in stock.

I also keep a supply of prepared, uncooked pork tenderloin. This is expensive, but when trimmed, sliced and battened out thinly, $\frac{3}{4}$ lb. is sufficient to feed four people. Some, I egg and crumb ready for frying, the rest is frozen in very thin slices, for a really spectacular dish – fry in butter, flame with brandy and reheat in single cream.

I rarely make cakes, because I don't like them. My husband, on the other hand, can be inveigled into doing anything, in return for a piece of chocolate cake. When I really need him on my side, I make a deep, 9-inch, rich, chocolate cake (5 eggs). The fudge type of icing is satisfactory for freezing and it sets quickly. When the icing has set, I cut and wrap the cake into individual helpings.

Basically, this is how I use my freezer. The one drawback as far as I'm concerned, is the difficulty of working within a budget. I've got as far as sticking a chart on the freezer cabinet and I usually manage to fill in the relevant details – that is, with the exception of the cost per portion, but this problem lies solely in a character flaw of mine.

On further consideration, I realise that convenience apart, being a freezer-owner must, inevitably, lead to economy. Whereas I used to spend 60p a head on the main course of a week-day evening meal, I now spend about half this amount which results in a weekly saving of £1·50 on this item alone. Of course, I also save considerably on entertaining costs."

Freezing on a shoestring

Some people may be keen to try home freezing but be loath to spend a lot of money on equipment only to find they can live without a freezer. (Personally, I can't see this happening. My experience is that most people, after a very short time, feel that their freezer has become a way of life and wish they had bought a bigger one in the first place!)

If you are nervous about a big expenditure on an untried and tested venture, why not look at the columns advertising second-hand goods? You can buy direct from the owner, or a renovated and guaranteed freezer from a firm which specialises in renovations. The guarantee usually covers only 1 month, but it is true that a freezer running properly during the first week will probably continue to function without needing repair for years. Most second-hand freezers come on the market because their owners are investing in a bigger model, not because the freezer is worn out. Even so it is best to take over any existing service facilities or ensure that they are part of the bargain you make with a dealer. Otherwise you should organise service facilities and consider insurance against loss of valuable food stock even if you feel the freezer itself may be a write-off if it goes wrong.

You must watch out that the bargain you are

condenser system are not so efficient in hot rooms, but are less prone to condensation and more suited to humid conditions. In a garage, mount it on wooden blocks to prevent rusting underneath.

The static machine type of freezer has power somewhere between the fan-assisted and skin-condenser types and makes little noise.

Some friends of mine decided to place the freezer in a large walk-in hanging cupboard in the bedroom, frequently unoccupied, adjoining their own. They found the noise very disturbing in the night, even through the bedroom wall.

Bulk buying at budget prices

It is only when you actually become a freezer owner that you can begin to take advantage of the wealth of opportunities to buy in bulk, and so begin to implement the claim that a freezer will save you money. You should find out what services are available in your area and then decide which will save you the most money, and will also be the best in terms of convenience and quality.

Do keep a watchful eye on the *quality* of the foods you can buy from the smaller freezer centres. Just because you can buy something a little cheaper, it doesn't mean that you are making the most sensible purchase. There is a lot of difference between the top grade frozen peas and the low grade ones; and it is false economy to have goods of inferior quality to save a few pence.

If you have a car available, you may decide to use a cash-and-carry freezer food centre. This is a collect-it-yourself service; your purchases will be packed into insulated bags, but from then on it is up to you to supply transport and get your goods home to your freezer before they thaw – probably about three hours, though in a heat wave it is less! Freezer centres (like Bejam) have a huge stock list to tempt you and often sell freezer accessories, too.

Other firms, like Brake Brothers of Tonbridge, will not only give a cash-and-carry service, but will deliver the goods if required.

The Birds Eye Home Freezer Service is a popular one and it is virtually nation wide. Telephone an order to your local depot and it is delivered to your home – Birds Eye products only,

offered doesn't turn out to be a conservator, able only to store food, not to freeze it down. Be careful, too, that the equipment is not so out of date that servicing may be difficult, spare parts expensive or unobtainable and satisfactory insurance cover dubious.

These reservations apart, second-hand freezers are likely to give excellent service. And maybe after several years' service you will feel you can pension off the old faithful and buy a big shiny new monster with all mod cons!

The question of noise

Noise affects us in different ways, so that while one freezer owner may be quite unperturbed by the sound of a freezer, her neighbour may shrink away from the noise.

Potential buyers must weigh up the pros and cons. If the freezer is to be sited in hot conditions, say in a glass surrounded conservatory, or in the kitchen near the cooker or boiler, a freezer with a fan-assisted cooling system is essential, but it is rather noisy – certainly noisier than the skin-condenser type. The freezers with a skin-of course, but these include the Smethurst, Tempo and Country Kitchen ranges.

Buying the meat should be treated in a slightly different way. A freezer centre may not be the best source simply because you purchase ready made up packs which will probably include some cuts of meat that you don't really want, or meat that is not butchered the way you prefer. Look for a wholesale butcher, like Price Brothers of Edgware, who specialise in supplying meat for the home freezer. And choose someone who not only offers competitive prices, but with whom you can discuss your exact requirements and preferences, and who will cut and package the meat in the most convenient way for you.

If there is a meat market nearby, you can buy direct from there. But meat bought this way may well not be portioned for you and, unless you are willing to tackle this none-too-easy job yourself, I should make quite certain on this point before you order – being presented with a quarter of beef to portion is no joke!

If you are fish eaters, and don't live too far from a fishing port or fish market, an occasional trip to buy some wholesale fish would be an investment. Make sure it is absolutely fresh, get it home as quickly as you can, and prepare and freeze it straight away because fish won't *stay* fresh long.

There are many firms throughout the country supplying a variety of services, but if you have difficulty in tracking down one that operates in your area, you should write to the **Food Freezer and Refrigerator Council, 25 North Row, London, W1R 1DJ** for their full list of suppliers, or to **The Electricity Council, Trafalgar Buildings, 1 Charing Cross, London, SW1A 2DS.**

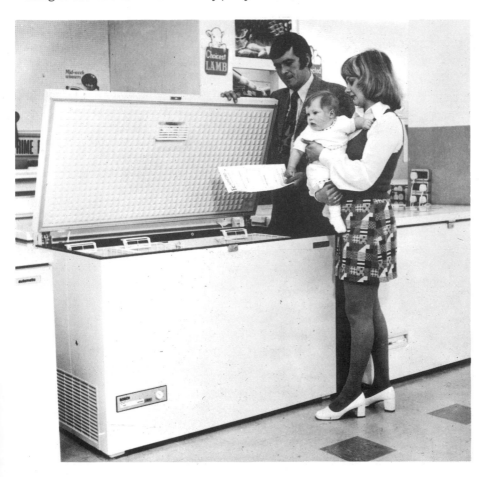

Chapter Three

Keeping your freezer running smoothly

All of us when we cast our minds back to school days can probably remember that warm air holds more moisture vapour than cold; and this simple fact plays the most significant part in frost formation in a home freezer. Warm air that enters the cabinet soon begins to cool, and when this happens, moisture is deposited in the form of frost. Every time the cabinet is opened, warm air flows in, bringing with it more moisture. It goes without saying, then, that frequent opening will increase the amount of frost that forms inside the freezer. Frost formation is more marked in the upright freezer because its front-opening door allows more warm air to flow in than does its chest brother. The mass of cold air in a chest freezer is trapped below the turbulent air at the top. This means that it is hard to displace the cold air, even if the lid is lifted frequently.

How to avoid frost formation

The amount of frost formed when the door is opened will depend on the surrounding temperature, and we wouldn't expect the same amount of frosting if the outside temperature were around 30–40 deg. F. rather than a good summer's range of 70–80 deg. F.

Although much frost formation would be avoided by opening the freezer seldom, this is not practical, so a certain amount of frosting is inevitable. Constant use of the freezer (and let's face it, that's the *best* way to use it) will cause some deposit of frost which is almost confined to the area near the opening and to a depth of about a foot all round the walls.

The frost in the cabinet itself will not seriously affect the efficiency of the freezer and de-frosting once or twice a year should be often enough – say when the frost is about $\frac{1}{2}$ inch thick. Even so, a certain amount of common sense should be exercised so that no unnecessary frost will be formed.

Hot foods

Hot, or even slightly warm foods, are often the culprits here. For not only do hot foods slow up the freezing process, they also cause extra frost deposit, throwing a heavy thermal load on the freezer. Before putting food into the freezer make quite sure that it is cold, by putting it in the larder – but not in the refrigerator while it is warm. Like most rules, the one about not putting hot or warm food in a home freezer, may be broken – but only in one circumstance. A freezer can be used to speed up chilling of a *very small* quantity of food which you want to serve ice cold – a fruit juice cocktail, for example, a milk pudding or jelly. Place the warm food (and the amounts *must* be small) well away from everything else in the freezer, so that it can do no harm.

Apart from the danger of excessive frost formation, there is another good reason why hot or warm foods shouldn't be put in the freezer. Many foods, and poultry in particular, actually suffer deterioration if they are frozen before they have been cooled sufficiently. In the case of poultry, it develops a noticeable green discoloration.

Large quantities

Large loads in the freezer also increase frost formation. For example, if a whole lamb was frozen down, it would increase the thermal load in the freezer causing more frost to form. Far better to joint the lamb, and make sure you don't overload the freeze capacity of your freezer. Put some packages of lamb into the freezer and the rest into the refrigerator until you can freeze them down also.

How to de-frost

To remove the frost from the lid and door it is necessary to scrape it off with a stiff bristle or nylon brush (not a wire brush), or use a wooden or plastic spatula. Avoid sharp tools that could damage the cabinet.

Once or twice a year the freezer must be de-frosted completely and cleaned. To do this, first turn off the current to the freezer, remove all the packages and store them carefully until de-frosting is completed.

Keep the lid of the freezer, or the door, open, to allow warm air to circulate.

Some freezers have a drain, but if yours has not, cover the floor of the cabinet with newspaper or a sheet of foam plastic, to catch the frost deposits so that they can be easily removed.

You may speed-up thawing time by placing tins and bowls of hot water in the freezer, replacing them when they have cooled to such a degree that they become ineffective – but, don't pour hot water directly into the cabinet, this amount of heat can create a great deal of pressure on the refrigerating system. On the other hand, *cold* water could be poured straight into the cabinet, for it helps speedy thawing without risking any damage, but it would only be effective in a freezer with a drain.

When the ice and frost has loosened from the sides, remove it, and when the cabinet is completely clear, clean it out. Dry thoroughly, close lid or shut door, and reconnect the electricity. One hour after the power is restored, the frozen packets may be returned to the cabinet; but the freezer won't be ready to freeze new packages for a further two hours.

Cleaning the inside

Make up a solution of bicarbonate of soda and warm water in the proportion of 1 tablespoon bicarbonate of soda to $\frac{1}{2}$ gallon of water. Wipe the inside of the cabinet with this solution. Don't use soap or detergents for they leave odours which may taint the contents of the freezer when it is refilled. Avoid caustic cleaners too for they will damage the surface. After the cabinet has been thoroughly wiped over with the bicarbonate of soda solution, rinse with clean water and dry. Clean the outside with white furniture cream.

Storing frozen food

While de-frosting is taking place, you will need to remove the frozen packages and store them in such a way that they will not begin to thaw. The easiest method of storage is simply to transfer the food to insulated bags and cover with a chilled blanket or rug. But if this isn't possible, spread a sheet of polythene, or brown paper, on the floor; pile the packages on top and cover with another sheet of polythene or brown paper. Finally, cover with a thick chilled blanket or rug to act as an insulator. When the power is off for longer than a few hours, you will need to take further action if the food is not to thaw and be wasted. In Chapter One, a section has been devoted to storing frozen foods in an emergency and in it you will find a number of ways this can be done. Don't however put dry ice *inside* the cabinet.

After de-frosting take the opportunity of bringing the inventory up to date.

Work out your budget to find your largest expense, seeing whether you can save by freezing cheap meats in cooked dishes, home-baked cakes or more fruit and vegetables.

Invest in a freezer thermometer to ensure that the temperature in the cabinet, at no time rises above 0 deg F. (the highest safe temperature for storing frozen foods).

Chapter Four

Preparing food for freezing

Now to get down to the actual processes of preparing and freezing food. To do this successfully, you must observe the golden rules carefully. Have a look at the following do's and don'ts of choosing and preparing food for freezing, so that you may avoid disappointment and waste of space in your freezer.

12 Golden Rules

1. Only top quality foods and ingredients should be used, for although freezing will retain the original quality, flavour, colour and virtually all the nutritive value of food, it does not, and cannot, improve it.
2. When you plan to freeze fruits and vegetables, buy or plant, for preference, the recommended varieties that are known to freeze well. Details of recommended varieties can be found further on in this chapter on pages 45 and 53.
3. It is vital to work quickly by preparing fresh produce as soon as it is received and hurrying the packages to the freezer before deterioration can take place. If there are too many to be frozen at one time, preserve the remaining packages in the refrigerator (or the coldest place possible) only until the freezer is able to cope with more unfrozen packages.
4. Plan your freezing programme ahead and make sure you have room for seasonal gluts, like strawberries.
5. Have adequate, high quality sealing and labelling materials on hand always. For example, special freezer tape, as ordinary adhesive tape will crack when reduced to a very low temperature.
6. Keep fresh meats and other produce chilled until they can be prepared for freezing – though for as short a time as possible. The longer fresh foods are kept at room temperature before freezing, the more colour, flavour and nutritive value they will lose.

7. Haphazard methods mean poor results, so make sure all hot or warm foods are chilled thoroughly before packaging, and cool them quickly, to minimise danger of bacterial growth. Work under hygienic conditions. Use a funnel when possible, to fill bags with prepared fruit and vegetables. This avoids unnecessary handling of the bag, and helps to keep the sides of the bag scrupulously clean.
8. Vegetables are meant to be blanched, not fully cooked, so follow the blanching timetable carefully.
9. Empty freezer space uses electricity, so keep your freezer fully stocked with foods, unless there is a reason for reducing stocks temporarily.
10. Materials and methods of packaging vary, so experiment with them. The criterion by which you judge any method or material is the satisfactory result you obtain from it.
11. Label packages carefully for quick identification and keep a date record of the food you put in and take out of the freezer.
12. All the food in your freezer will eventually deteriorate, so remember to use it up while it is still in excellent condition. Don't forget to use up the bottom layer of packs in rotation with the upper layer.

Glossary of terms
There are a number of terms used in connection with home freezing, and although they are described fully in appropriate sections of this book, the glossary here will give you a general idea of their meanings.

Anti-oxidant
Chemical agent, such as ascorbic acid, added to sugar or syrup to control fruit discoloration.

Ascorbic acid
Synthetic Vitamin C product available from chemists. Use $\frac{1}{4}$ teaspoon acid to $\frac{1}{2}$ pint water.

Blanch
To heat vegetables in boiling water, or to steam just long enough to slow or stop enzyme action.

Butcher's wrap
Method of wrapping food suitable for freezing. Place food near corner of paper, fold corner over food, then fold two sides across top. Roll the package over so that wrapping encircles food. Fasten with freezer tape.

Druggist's wrap
Method of wrapping food suitable for freezing, in which it is placed on centre of paper. The two horizontal ends are brought together and folded over and over until tight against the food. Seal, then tightly fold in the two end pieces and seal.

Dry pack
To pack fruits without adding liquid or sugar.

Dry sugar pack
To pack fruits in dry sugar.

Enzymes
Chemical agents which ripen vegetables and fruit.

Freezer burn
Dehydration of badly wrapped food. It is caused by oxidation. Loss of colour, quality, flavour and texture are often the result.

Glaze
To dip frozen food quickly into water which is only just above freezing point. A thin coat of ice forms over food.

Headspace
Space left at top of container to provide room for expansion on freezing.

Heat seal
To seal a package wrapped in polythene, with pressure from a warm iron or electric heat sealer.

Moisture-vapour-proof
Freezer packaging materials that are treated to prevent moisture and vapour evaporation.

Stockinette
A tubular mesh material usually used for outer wrapping meat.

Syrup pack
To pack fruits in sugar syrup of varying strengths.

Why speed is necessary
From the first moment that the food begins its journey to the freezer, whether this journey begins on a fruit tree, in an allotment garden or the farmyard, speed is of the very essence. Fruit and vegetables should be packed at the peak of their perfection, ripe but not too ripe; firm but tender and juicy. Choose the right moment for picking, and make sure you have planned your freezer space to cope with the seasonal batches of food. Once picked, see that the prepared and packaged foods are stowed away in the freezer at the earliest possible moment – within an hour or two, if possible. The same principles of top quality and quick preparation apply to meat and, indeed, to any other kind of frozen foods. An item that is held for any length of time before freezing takes place, is liable to considerable deterioration in quality and the danger from bacterial growth is increased.

The opposite rule applies to thawing. Slow thawing in the refrigerator produces better results than quick thawing at room temperature, which sometimes causes damaging changes in the structure of the food. The danger is also avoided of raising the temperature of the food above that of the refrigerator cabinet – into the temperature range above 40 deg. F. – when bacterial growth becomes much more active.

The importance of Vitamin C
Most of us know that extreme care must be taken in the cooking of fruit and vegetables so that the loss of Vitamin C is kept to the minimum. Exactly the same care must be taken in the preparation of fruit and vegetables that are to be frozen, for without it, the loss of Vitamin C can be considerable.

Vitamin C, otherwise known as ascorbic acid, is found in cells of plants. In these cells are also enzymes. When the cells are ruptured by cooking or chopping, the chemical action of the enzymes causes loss of Vitamin C content. Vitamins form a valuable part of our diet. Whereas most other nutrients are available in many foods so that if you miss them in one form you pick them up in another, Vitamin C is much more elusive – the chief source being fruit and vegetables. The amount they contain quickly diminishes, either through bad preparation and cooking, or simply old age!

This is why the blanching times recommended in the Vegetable Preparation section (see page 43) should be so strictly adhered to. These times have been worked out scrupulously so that, while the harmful enzyme action is slowed down, the Vitamin C loss is kept to the very minimum – in other words, if the blanching instructions are followed exactly, the Vitamin C content will not reduce by more than a quarter. Even when a reduction of Vitamin C occurs, it is possible to make up the deficiency to some extent with ascorbic acid tablets. These are also useful in helping certain fruits to retain their colour, but this aspect of their use is discussed in the section on fruit.

Most Vitamin C loss occurs during preparation and cooking of food. Very little is lost during the actual freezing and you can expect to conserve approximately three-quarters of the vitamin content after six months at 0 deg. F., or lower.

What are headspaces?

When food is packed for freezing it is usually necessary to leave a space between the top of the food and the lid of the container, to allow for expansion caused by putting food into sub-zero temperatures. Dry packs need about $\frac{1}{2}$ inch, wet packs in narrow topped containers need $\frac{3}{4}$–1 inch, wide topped containers can have $\frac{1}{2}$–1 inch. If larger than pint-size containers are used, double headspace is needed.

How to avoid spoilage

The natural process of food spoilage is caused by the activity of micro-organisms – that is bacteria, moulds, yeasts – and chemical agents known as enzymes. The only way to slow down this activity is to store foods in a sufficiently low temperature. Few bacteria grow below a temperature of 40 deg. F. (4 deg. C.) – the usual temperature of a refrigerator. Yeasts and most moulds stop growth below 14 deg. F. (– 10 deg. C.). Enzyme activity can never be stopped entirely, but low temperatures slow down this activity so that it becomes harmless while food is stored in a freezer.

Contamination

The best defence against contamination is careful packing, freezing and thawing. Indeed, if the work is done scrupulously, contamination will not take place. It is usually caused by dirt, moulds, cross-flavours and insects. Food must be enclosed in a satisfactory packing material to withstand very low temperatures for periods of up to a year. All possible air must be pressed out of the package before sealing – for air aids dehydration and cross-contamination. With air in the pack, fat on meat, for example, can turn rancid sooner and oxidation may take place (see page 64) causing 'freezer burn'. Dehydration and consequent 'freezer burn' are non-toxic but they produce unappetizing results.

At temperatures below 18 deg. F. organisms of any kind remain dormant and cannot be transferred from one package to another if this temperature is maintained. Odours and flavours will contaminate other foods at any temperature. It is, therefore, vital that all food should be carefully wrapped before storing.

The freezing itself must be done quickly so that the food drops below the temperature range which aids the growth of harmful germs and bacteria with the least possible delay.

The safest way to thaw food is in the refrigerator, though this inevitably takes longer than thawing at room temperature. Once thawed, it must be cooked and eaten as soon as possible. When food rises over the safe temperature level, bacteria moulds, yeasts and enzymes reactivate.

Special points to watch out for

1. Either wipe clean, dry and freeze whole or wash and blanch mushrooms in the normal way so that dirt and germs are removed. Be sure they are fresh and completely free from decay before freezing.
2. Take special care in cleaning such fruits as blackberries and raspberries. Small insects can lurk unseen in crevices, increasing the dangers of contamination.
3. Only pasteurised or double cream with over 40% butterfat content should be frozen, including Cornish and Devonshire clotted creams which have been treated with heat.
4. Game birds and animals should be hung before freezing. Thawed game goes bad quickly.

28

Unsweetened thick apple purée can be frozen in the smallest size Tupperware containers, to serve as apple sauce with roast pork or other savoury dishes.

The remainder of the same batch of apple purée can be sweetened and packed in used yogurt or cream cartons for baby-size portions of apple dessert.

Small portions of onion sauce to serve with lamb can also be packed in individual containers. Bread sauce to go with poultry also freezes well.

Other cooked fruit purées make lovely puddings for babies. Apple and black-berry, or apple and blackcurrant are particularly successful.

5. It is generally safer to thaw food in the refrigerator but this is especially true of pork.

6. It is better to store meat and poultry without stuffings. Although stuffing may be frozen successfully, its freezer-life is less than that of the meat or poultry.

7. When freezing a casserole, avoid the troubles air can cause, by inserting a knife through the ingredients to released trapped air, then cover.

8. For safe, even de-frosting, thaw cold, sliced meat in the refrigerator if serving it cold. But if you are serving it hot, reheat frozen or partially thawed meat thoroughly and quickly in gravy, sauce or a cooked mixture.

9. When buying commercially frozen food, transfer to your freezer at home as quickly as possible – at the most, three hours if using an insulated bag. Ideally one would buy frozen food from a cabinet marked at its load line 'B.S. 3053:1971' which incorporates all the standards of performance required for shop freezers. This does not mean that freezers without this mark do not freeze satisfactorily, just that, with it, you can expect the very best performance. Make sure all packages are stored below the load line.

10. Don't re-freeze food once it is completely thawed. If you are uncertain whether to re-freeze check with the chart on page 79.

11. Always make a note of the method you have used because you may want to use this method again or amend it in some way.

12. Don't leave any pockets of air inside the

packages. These collect moisture from the food making it dry out.

13. When adding food to the freezer, allow air spaces between packets to ensure quick freezing. Close packing delays freezing.

14. Keep foods of a similar type together in the freezer so that you know where to look for it. The use of coloured nylon string bags or polythene batching bags for this purpose is a great help.

15. Don't spoil made-up dishes such as casseroles, by adding ingredients that don't freeze well. Some sweet herbs come into this category.

Soups

Soups are simplicity itself to freeze and almost all types are frozen successfully with the exception of those with a milk or cream base. This problem can be eliminated to a large extent by adding vegetable purées to freshly made pouring sauce (1 oz. butter or margarine and 1 oz. flour to each pint of milk) to make time-saving vegetable soups.

Potatoes, rice, barley and other starchy ingredients in soup recipes are better avoided. Make the soup in the usual way, then strain, cool and remove surplus fat. Pour into rigid polythene containers leaving $\frac{1}{2}$–1 inch headspace if a wide-topped 1-pint container is used. A narrow topped container should have $\frac{3}{4}$–1 inch and for a larger container double headspace is needed.

Sauces

Keeping an assortment of basic sauces in your freezer opens a whole world of exciting cookery. Freeze several portions of basic savoury white and brown sauces. And make sure you have a supply of a rich tomato sauce as well.

It is simplicity itself to reheat the sauce of your choice, or add to it the special ingredients you need for a more complicated dish. Sauce can be frozen with cheese in it already, but use processed cheese to give a smoother texture when de-frosted.

To pack, use a rigid based polythene container leaving a headspace which depends on the size of the container, and cover with an air-tight seal. Alternatively, freeze heat-sealed in boiling bags, so that they can be immersed in boiling water and simmered for 10 minutes to heat for serving.

It is better always to add cream or soured cream when the sauce is de-frosted and reheated, as these ingredients tend to separate on freezing. Always avoid ingredients that don't freeze well.

FISH

The deterioration rate of fish is very high and it is essential to freeze it as soon after catching as possible. In fact, freezing fish is only practicable for those living near fishing facilities since it must be prepared and put in the freezer within twenty-four hours.

Fresh fish can be recognised by its bright colours, clear, bright eyes, red gills and firm elastic flesh. Anglers should not allow the fish to bounce around in the boat or on the bank, or they will become bruised. Kill quickly after catching, and remove internal organs and gills straight away. Keep cool in an insulated bag until it can be got home.

The lean fish – such as haddock, cod and halibut – have better keeping qualities than oily fish. They can be kept for six months while oily fish – sole, herring, mackerel, etc. – can only be stored for three months.

How to prepare for freezing

The fish should be prepared as if it were being cooked. Remove the fins, tail and then scale. Rinse in cold water. Fish can be frozen whole, filleted or cut into steaks. If it is to be cut up do this now, then dip lean fish in a salt and water solution (2 oz. salt to 2 pints water) for 20 seconds. Oily fish don't need the salt and water dip, but if they have been cut, put them in an ascorbic acid solution (2 teaspoons ascorbic acid in 2 pints water) for about 20 seconds. Put enough fish in one package to serve a family-sized meal. Separate the pieces of fish with freezer paper, then wrap tightly, excluding air. Avoid fishy smells by overwrapping with brown or grease-proof paper.

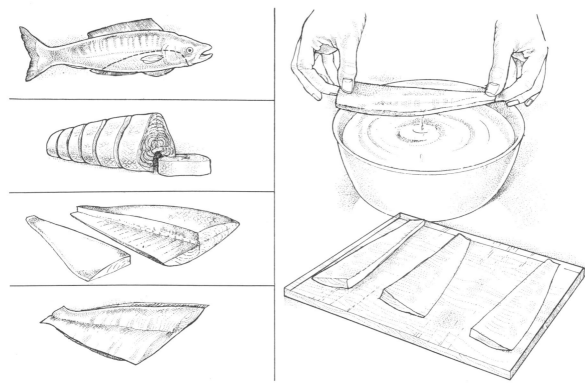

Glazing

Give large, whole fish extra protection by glazing it before wrapping. To do this, open freeze fish until solid then dip it quickly into ice-cold water. A thin coating of ice will freeze on the surface. Repeat this process several times, then wrap the fish in freezer paper. Seal, label and freeze, store for up to 3 months.

Shellfish

If anything, shellfish are even more perishable than any other fish so if you have any doubt whatever about the freshness of the shellfish, *don't freeze*. Clean, cook or prepare as if they were to be eaten immediately. Shell cooked shellfish before packing. Store for up to 1 month.

Individual instructions for preparing fish

Whole fish

Some fish such as herring or mackerel are suitable for freezing whole. First remove the scales with a dull bladed knife, or special scaling tool. Then remove the head, backbone, fins and tail and gut.

Glaze if liked, then wrap tightly, seal, label and freeze. If moulded in foil, cook in the pack.

Fillets

Cut the fish into fillets, then remove the skin by putting the fish on a chopping board, skin down. Using a sharp knife, insert the point between skin and flesh at the tail end, rub with kitchen salt and, holding the knife steadily, pull the skin towards you allowing the knife to move slowly forward. Keep the knife close to the skin but don't cut through it. Soak oily fish in an ascorbic acid solution for about 20 seconds. Lean fish should be dipped in a salt and water solution for 20 seconds. Wrap tightly, seal, label and freeze.

Salmon

To freeze a whole salmon, de-scale with a blunt knife. Remove the internal organs, gills and fins. Wrap closely in sheet polythene, moisture-vapour-proof paper, freezer foil, or a double thickness of kitchen foil.

Unless storing salmon for a special party, it is more practical to freeze it in portions.

31

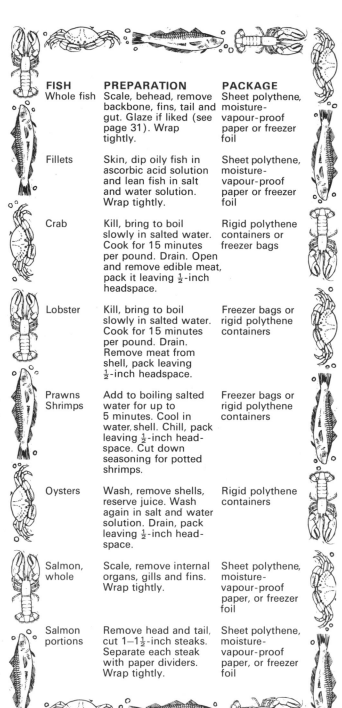

FISH	PREPARATION	PACKAGE
Whole fish	Scale, behead, remove backbone, fins, tail and gut. Glaze if liked (see page 31). Wrap tightly.	Sheet polythene, moisture-vapour-proof paper or freezer foil
Fillets	Skin, dip oily fish in ascorbic acid solution and lean fish in salt and water solution. Wrap tightly.	Sheet polythene, moisture-vapour-proof paper or freezer foil
Crab	Kill, bring to boil slowly in salted water. Cook for 15 minutes per pound. Drain. Open and remove edible meat, pack it leaving $\frac{1}{2}$-inch headspace.	Rigid polythene containers or freezer bags
Lobster	Kill, bring to boil slowly in salted water. Cook for 15 minutes per pound. Drain. Remove meat from shell, pack leaving $\frac{1}{2}$-inch headspace.	Freezer bags or rigid polythene containers
Prawns Shrimps	Add to boiling salted water for up to 5 minutes. Cool in water, shell. Chill, pack leaving $\frac{1}{2}$-inch headspace. Cut down seasoning for potted shrimps.	Freezer bags or rigid polythene containers
Oysters	Wash, remove shells, reserve juice. Wash again in salt and water solution. Drain, pack leaving $\frac{1}{2}$-inch headspace.	Rigid polythene containers
Salmon, whole	Scale, remove internal organs, gills and fins. Wrap tightly.	Sheet polythene, moisture-vapour-proof paper, or freezer foil
Salmon portions	Remove head and tail, cut 1–$1\frac{1}{2}$-inch steaks. Separate each steak with paper dividers. Wrap tightly.	Sheet polythene, moisture-vapour-proof paper, or freezer foil

Remove the head and tail, then cut 1–$1\frac{1}{2}$-inch steaks from both ends. The middle, and thickest, part of the fish can be cut into family-sized pieces, to poach or bake on the bone. Stack cutlets with dividing papers between them, before wrapping for freezing. Store for up to 3 months.

Crab

First kill the live crab, put in salted water and bring to the boil slowly. Estimate the cooking time by allowing 15 minutes per pound. Drain. Open up and remove the edible meat from the body and claws, pack it in bags or polythene containers leaving $\frac{1}{2}$-inch headspace. Seal, label, freeze. (Keep the cleaned shell for serving.)

Lobster

Follow the same cooking method as for crab. When cooked, remove the meat from the shell, pack it in bags or polythene containers leaving $1\frac{1}{2}$-inch headspace. Seal, label and freeze.

Prawns, shrimps

Put into boiling salted water for up to 5 minutes (a little less for small shrimps or prawns). Allow them to cool in the water, shell. Chill, pack into bags or polythene containers leaving $1\frac{1}{2}$-inch headspace. Seal, label and freeze. Potted shrimps can be frozen, but season sparingly, as they taste salty when thawed.

Oysters

Wash thoroughly and remove shells reserving the juice. Wash again in a cold salt and water solution, drain and pack in polythene containers adding the reserved juice. Leave $\frac{1}{2}$-inch headspace, seal, label and freeze.

MEAT

The advantages of freezing meat at home are many and varied. With meat tucked in your freezer you are never at a loss when unexpected visitors drop in. You have always the basis of a good staple meal if the weather is bad and you can't get out shopping; or have a sick child on your hands, when catering is the last thing you want to bother about.

Probably one of the most important advantages of freezing meat is the great saving on the family budget that can be achieved.

Before having a freezer, few housewives know about buying wholesale meat. The charts opposite are designed to help. The best cuts of beef are found in the hindquarter (weighing about 160 lb.) while the forequarter (about 100 lb.) provides the mincing, braising and stewing beef. A whole lamb weighs about 45 lb. for English, 26 lb. for New Zealand – or buy just half a lamb. Pigs can weigh up to 100 lb. but, like lamb, are sold halved. Make it clear whether you want the head and trotters.

Meat can be bought in bulk at greatly reduced prices. A whole lamb can be jointed in the usual way, then frozen, providing leg joints, shoulder, loin and so on. The cheaper bony cuts of lamb usually sold for stewing take up too much freezer space if left as they are. It is better to make them into stews and remove the bones before freezing. A quarter of beef is the most usual amount bought in bulk, but this often yields far too much meat for one family for it can weigh 200 lb. See if a friend with a freezer will go halves with you in buying a quarter of beef, then divide it up in the way most satisfactory to both. Alternatively, buy the beef already jointed when it is easier to select only the amount you need to freeze for yourself. A half pig provides a reasonable amount of meat for a family, allowing for the fact that you won't want to eat pork every day for 3 months! It is therefore a good buy.

Remember that if you are asking your butcher or wholesaler to joint the meat for you, you must expect a small charge for doing it. Indeed, it is a good idea to indicate that you expect him to charge you, for the poor man has to make a living like anyone else!

Specialist frozen food suppliers sell parcels of jointed and prepared cuts of meat in bulk. If you have a smaller freezer and yet still want to buy meat at a reduced price – and who doesn't? – a neighbourly syndicate may well be formed in order to buy meat in this way. Agree beforehand how much of the animal each person will be buying and divide up the cost accordingly. Even a neighbour without a freezer would welcome a

BEEF
Forequarter

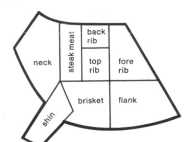

BEEF
Hindquarter

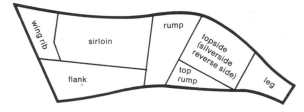

LAMB

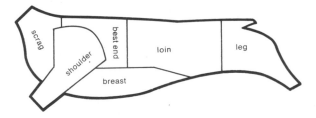

PORK

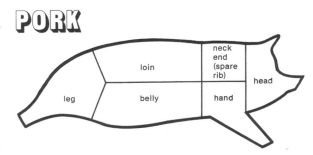

33

Wrapping a joint of beef in
moisture-vapour-proof material.

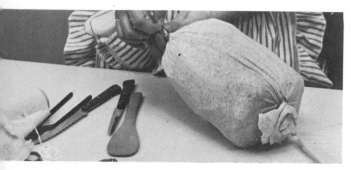

Over-wrapping with stockinette to
preserve pack from damage.

Cut a piece of polythene film large
enough to allow a 3–4 inch overlap
around a shoulder of lamb plus
2–3-inches at both ends. Place lamb
in centre and fold the two longest
edges together and fold them over
and over until they are tight against
the meat. Fold in ends and secure.

piece of topside for Sunday if she could get it at a
bargain price, so pass the word around.

Choosing meat

As with any other product that is to be frozen,
high quality meat must be chosen. The freezing
process will maintain the quality and freshness of
a joint, but it cannot improve one that was poor
quality to start with. As long as the meat is
prepared, frozen and thawed in the recommended
way, it will have practically the same goodness as

fresh meat. If you short-cut the preparation, and
especially the thawing, the quality of the meat
suffers.

Boning the meat

Meat is a bulky product and valuable freezer
space can be lost if you freeze the bones as well as
the meat. A better plan is to bone the meat
whenever possible, for as well as increasing the
space in the freezer cabinet, it will reduce the
common danger of bones tearing the outer
wrapping, allowing air to creep into the package.
Don't discard the bones, but make up well-
reduced soup stocks from them. Allow them to
cool, then pour into containers, seal, label and
freeze in the usual way. If you are short of containers, freezer bags can be used for stock, or any
other liquid (see page 65).

Packing

The material used for packing meat should keep
the air out of the packages and the moisture and
flavour in. Pack in sizes suitable for your family;
it is pointless freezing packs of meat so large that
you have too much to use up before the joint
starts to go off. The leftovers cannot be re-frozen,
and the beautifully preserved meat then suffers
deterioration.

Unboned meat

It is not always practicable or possible to bone
certain cuts of meat, and if any of the bones are
protruding they must be covered carefully with a
pad of aluminium foil to stop them piercing the
outer wrapping. If air is able to penetrate the
wrapping, the meat suffers considerably; it
dehydrates and becomes a tasteless, sad-looking
joint compared with the tempting, juicy, red meat
that usually emerges from a freezer. The dangers
from cross-contamination are also much increased.

It is a good idea to cover up the whole wrapped
joint with a piece of stockinette or an old clean
stocking to provide extra protection from the
bones. The photographs on wrapping poultry
(see page 41) show how this is done.

Storage time-table for meat

If meat is selected carefully, packed properly and frozen and stored at sub-zero temperatures, it will keep satisfactorily for the following periods of time:

Meat	Months
Beef	9–12
Pork, chops	3–4
Pork, roasts	4–6
Veal, escalopes, chops	3–4
Veal, roasts	4–6
Lamb, chops	3–4
Lamb, joints	9–12
Minced meat	1
Offal	2–4
Cooked meats	4–6
Sausages, seasoned	1
Sausages, unseasoned	6

General principles of freezing meat

Meat for the freezer must be thoroughly chilled as soon after slaughter as possible. Beef and lamb are easiest to cope with, in that there is no extreme urgency in getting the meat into the freezer. As long as it is kept in cool hygienic conditions, it will be all right for up to 7–10 days after slaughtering. Veal and pork must be handled much more quickly and should be safe in the freezer within 5 days of slaughtering.

Don't freeze more meat than you can eat. In other words, meat can only be stored for a certain length of time, as you will see from our storage time-table. Remember that you will only want to eat a particular type of meat occasionally, so don't take up freezer space with more meat than you want.

Try not to prepare more meat than you can freeze in one batch.

Individual portions, such as chops, fillets and steaks, can be packed together each separated by a sheet of moisture-vapour-proof paper. This makes it easy to separate them before cooking – a boon if you don't want to wait for complete thawing to take place. Don't pack too many together in one package, though – just enough for the usual number of people you are likely to serve at one meal.

Before packing, trim off all the surplus fat. Fat reduces the storage time because it becomes rancid much more quickly than the meat. Hard fats keep better than the soft ones. The fat is the first to suffer if the meat is held too long before freezing; it loses flavour, becoming quite unpalatable, and increasing the chance of rancidity. Even so, don't remove all the fat because it prevents the meat from drying out and adds greatly to the flavour.

Individual instructions for preparing meat

Meat joints (Beef, lamb, pork and veal)

Whenever possible, remove the bones. Cut away surplus fat leaving a little to add flavour to the joint. Roll and tie the joint into shape, wipe with a clean cloth. Cover any remaining sharp bones with a pad of foil to stop them piercing the outer wrapping. To wrap the bones, tear off a fairly large piece of foil, big enough for you to fold it over two or three times to form a 2–3-inch wide strip. Wind the strip of foil around the protruding bone making sure all the rough edges are covered. Wrap in a polythene bag, excluding as much air as possible, seal, label and freeze.

Minced meat

This is best frozen without salt. If you are mincing the meat yourself (which is the best way), trim off surplus fat first, then pack the minced meat tightly into a polythene bag or polythene container making sure there are no pockets of air. Seal, label and freeze.

Offal (Heart, kidney, liver, sweetbreads, and tongue)

Offal must be prepared and frozen quickly. Liver is the most successfully frozen offal, though it shouldn't be kept for more than about 3 months. Tongue, heart and kidneys can be kept for a little longer, up to about 4 months. Wash offal thoroughly before freezing and remove the blood vessels. Dry carefully, then wrap in polythene and place in a polythene bag or rigid polythene container. Seal, label and freeze.

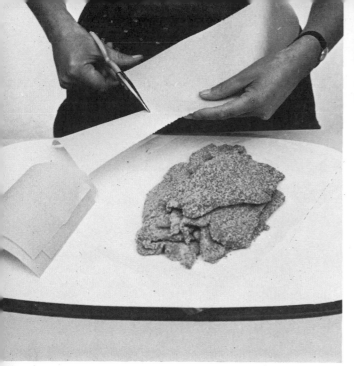

Veal escalopes can be frozen ready-coated in egg and breadcrumbs. Use freezer paper or foil as dividers so that they can easily be separated.

Stack the escalopes on top of one another, separating each one with a paper divider. Alternatively, the dividers can be made from foil.

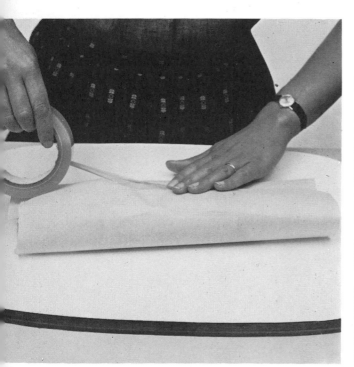

Fold the second edge over the first and seal with a strip of freezer tape. Press ends down closely against sides of meat.

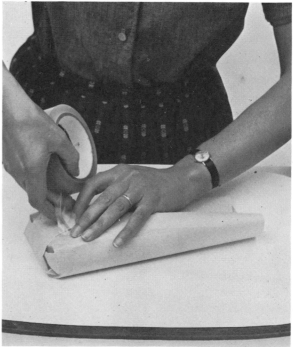

Now fold the open ends parcel-fashion and seal with more freezer tape, making sure you have a neat, air-tight package.

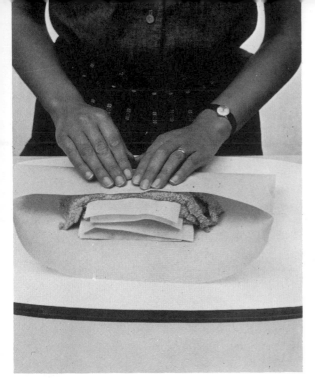

Cut a piece of freezer paper large enough to make a neat parcel with proper tuck-in flaps at each end. Place the veal escalopes in the centre and fold one edge over the escalopes.

Label the pack clearly and freeze. Alternatively, put several unsealed packs into a rigid-based polythene container with air-tight seal.

Stewing meat

Trim off the surplus fat, cut meat into 1-inch cubes, then pack tightly into a polythene bag or plastic container, excluding all pockets of air. Seal, label and freeze.

Chops, cutlets, steaks

Estimate the number of portions required for one family meal, and pack them together. Each chop or steak must be divided with a piece of foil or freezer paper to make separation easier when

Before wrapping the chops, prepare a number of foil dividers large enough to keep all the chops separate.

Place dividers between all the chops so that they can easily be separated when removed from the freezer even though they have not thawed.

Transfer the chops to the centre of a piece of moisture-vapour-proof paper, cut large enough to provide a good overlap, plus extra at the ends.

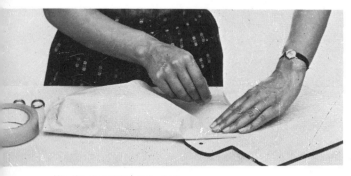

Use the druggist's method of wrapping and fold the end pieces neatly to help ensure a good seal.

the package is removed from the freezer. Press out air by packing together tightly in polythene bags or plastic containers. Seal, label and freeze.

Sausages

Shop-bought sausages are unsuitable for long freezing because they have a high fat content and contain bread and seasoning. This reduces the storage time of sausages to about 1 month. If you make your own, keep seasoning to the very minimum if you want to preserve them a little longer. Home made sausage meat that is unseasoned can be kept for up to 6 months.

Ready-cooked dishes

Casserole dishes and stews are particularly suitable for freezing, and even if the meal was made with frozen raw meat, it can still be frozen again once it is cooked – one of the few occasions when it is possible to re-freeze meat. A word of warning – only the two freezings are possible, once when the meat is raw and once when it has been cooked. Don't be tempted to re-freeze a cooked casserole dish once it has thawed.

Not all traditional stew and casserole ingredi-

ents are suitable for freezing and it is better that these are added after freezing at the reheating stage. These ingredients include potatoes, rice, spaghetti, macaroni and any other kinds of pasta. Seasoned food becomes stronger when it is frozen, so either omit the seasoning till the dish is reheated, or keep it to the very minimum.

This is the type of food which lends itself to the 'eat one, freeze two' principle. So prepare plenty at one time – at least enough for three separate family-sized meals. To freeze: carefully line a casserole dish with freezer foil, and then pour in the chilled stew, cover with another sheet of foil and put into freezer until the food is frozen, but not too hard. Remove from the freezer and lift the foil lining out of the dish. Cut the food into family-sized meals and wrap individually. This makes it easy to take out just one. (See the step-by-step photographs on pages 38–40.) If you freeze all the stew in one container, take care to use a casserole dish which is suitable for sub-zero temperatures. Pyrosil, for example, will survive very low temperatures, and can be taken straight from the freezer and put into the oven; but this is a rare quality and few other dishes are suitable for this treatment.

Cut a piece of foil large enough to line the casserole plus an extra 2 inches all round. Press the foil into the casserole, leaving overlap to allow easy lifting when the foil and stew are removed.

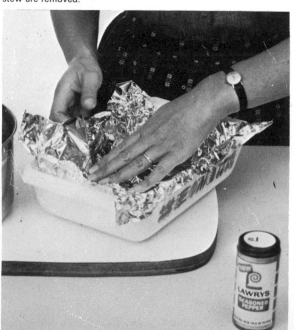

Put stew into the lined casserole dish, spreading it evenly into each corner, and smoothing over the top.

Cut another piece of foil 2 inches longer and 2 inches wider than the top of the casserole. Place it on top.

Twist the edges of the foil around the dish, lodging the edges under lip-edge if there is one. Put into the freezer.

When partially frozen remove from the freezer, uncover and carefully lift the stew out of the dish.

Using a sharp knife, cut the stew into three even-sized bricks. Cut pieces of foil, each large enough for one portion.

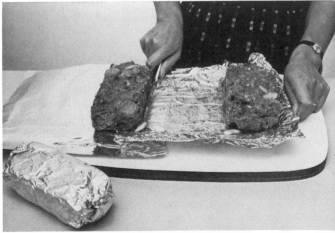

Slide a brick of stew into the centre of a piece of foil. (Freezer foil or ordinary kitchen foil used double thickness.) Take care not to tear foil.

Hold both ends of the foil above the centre of the stew. Fold them over and over, druggist-fashion, until the seam is flat against the food.

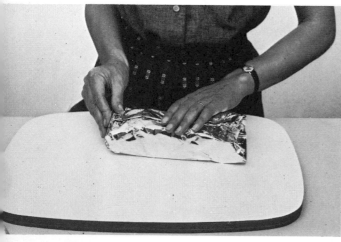

Smooth the flap up the side, to meet the centre seam. Close the other end of the parcel in the same way.

Label and freeze. There is no need to use freezer tape on well-wrapped foil packages, they are self-sealing.

POULTRY

Poultry is just as versatile as meat, in that it can be frozen fresh, cooked, or in made-up dishes. It can even be frozen in sandwiches. Select fleshy birds with well-distributed fat and, in the case of oven-ready birds, look for those with few skin blemishes. If you buy a bird wrapped in film or in a display pack, remove this wrapping before freezing and cover with a material suitable for freezer storage. When you are buying poultry for immediate use, choose the bird according to the way you want to cook it. Young birds for roasting, frying and grilling; more mature birds for braising and stewing. Bear this in mind when selecting poultry for the freezer and mark on the pack the intended use. Oven-ready birds need little preparation for freezing. Simply remove any pin-feathers or lung tissue that may have been left inside, then wash the bird inside and out. Don't stuff the bird before freezing. Stuffing increases the length of freezing and thawing times, risking the growth of harmful bacteria. Also stuffing contains seasonings which would reduce the keeping qualities of the poultry.

Individual instructions for freezing poultry

Whole birds

The term 'birds' applies equally to chickens, ducks, geese and turkeys – and all of them may be frozen whole and ready to roast. First, clean the bird, then tie the legs together with string or a rubber band, then press the wings close to the body. Cover protruding leg bones with a pad of foil to prevent the outer wrapping from being punctured. Alternatively, wrap them with grease-proof paper and secure with a rubber band. Freeze the giblets with the bird only if it is to be eaten within 3 months. Birds to be frozen for longer periods should have their giblets frozen separately. Giblets frozen with the bird should first be wrapped in polythene or other suitable material, then placed in the body. Put the bird in a large freezer bag or wrap in another suitable freezer material, seal, label and freeze. (See the photographs opposite which show how to wrap a chicken for freezing.) Giblets that are frozen separately can be put into a suitable bag, sealed, labelled and frozen.

Half birds

It is sometimes more convenient to divide large turkeys and chickens in half, before freezing. To do this, lay the bird on one side and cut from neck to tail along both sides of the back bone. It is then possible to remove the neck and back strip. Then, lay the bird on its breast, open it and cut along the inside of the breastbone. The halves may be packed together as if they were still a whole bird, or they can be packed separately. When the halves are frozen together, divide them with a piece of freezer paper. The giblets can be dealt with in the way described on the left. Put the

half or halves in a suitable bag or wrap in freezer material, seal, label and freeze.

Chicken or turkey portions

If the poultry is to be fried, fricasséed, stewed or casseroled, it is more economical on space if it is jointed before freezing.

After jointing, wash the portions in cold water and dry. (Don't freeze the bony pieces such as neck, back or wing-tips. Use these to make chicken or turkey broth. When prepared, pour into rigid polythene containers, seal, label and freeze.) Wrap the chicken portions in suitable material, seal, label and freeze. The step-by-step photographs on pages 42–43 show another way of packing chicken portions, using foil containers.

Sharp bones can pierce wrappings and allow air to creep in. Protect food by padding bones. Take a large piece of foil and fold several times to form a thick padded strip.

Carefully wind the strip of foil round the protruding bone. Make sure that no sharp edges are left unpadded.

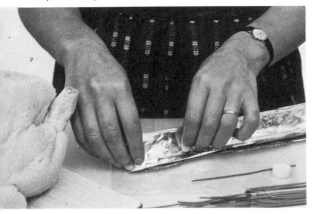

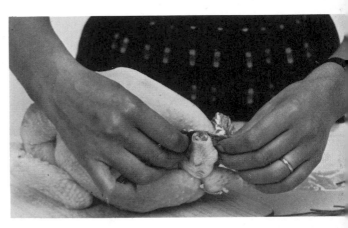

Now place in a polythene bag, press out all the air and twist the neck of the bag over several times. For poultry and large joints you will need a gusseted bag so that they will fit snugly without straining the polythene.

To complete the seal, wind a plastic covered wire fastener round the bag tightly. Label clearly — the small adhesive labels are just the thing for polythene packages. The usual freezer pens and pencils aren't satisfactory on polythene.

Aluminium foil trays are useful for holding joints of chicken. Divide the chicken into convenient size portions and put one portion in each tray.

Prepare pieces of freezer foil, large enough to cover the top of the tray with an extra inch all round to allow sealing.

Cooked poultry

Remove the meat from the bones and cut into small slices. Pack tightly into polythene containers and cover with broth or gravy, leaving $\frac{1}{2}$-inch headspace. Seal, label and freeze. Cooked chicken is not suitable for freezing if it is to be served cold.

Stews and casseroles

Dishes of this kind, made with poultry, are especially suitable for freezing. Only add potatoes, rice, macaroni and spaghetti in small quantities or omit them from the dish and add them at the reheating stage. Turn to page 38 for instructions on how to freeze and see how this is done in the step-by-step photographs.

GAME

The luxury of out-of-season game is one of the chief appeals of freezing it, but some extra care in preparation is needed. Game birds should be bled as soon as they have been killed and should be kept cool until you can get them home. Game animals should also be bled soon after killing, and they must be beheaded at this time. Game must be hung before freezing. If you try to do it after thawing, the flesh will go bad. The length of hanging is a matter of individual taste – and also depends on the weather. Some people like it hung until it acquires that strong gamey flavour which is so distinctive; others prefer to eat it sooner. Game should be hung longer in cold, frosty, weather than when it is damp and warm. In any case, if you are home freezing game, hang it for one day less than the usual period. This is because the bird or animal matures still further

MEAT Beef	PREPARATION	PACKAGE
Joints	Either bone or cover bone with foil. Roll and tie into shape if boned. Wipe with clean cloth. Wrap tightly.	Moisture-vapour-proof paper or freezer foil. Over-wrap with stockinette.
Steaks	Trim off fat. Separate with foil or moisture-vapour-proof paper dividers. Wrap tightly.	Moisture-vapour-proof paper or freezer foil over-wrapped with polythene.
Stewing steak	Divide into one-meal-sized portions. Place in freezer bags.	Freezer bags, sealed with ties or by heat.
Mince	Use very fresh. Omit seasoning. Divide into portions.	Freezer bags, sealed by ties or by heat.
Lamb Joints	See beef joints.	See beef joints.
Chops and Cutlets	Stack with foil or moisture-vapour-proof paper dividers. Wrap tightly.	Moisture-vapour-proof paper or freezer foil. Over-wrap with stockinette for extra protection.
Pork Joints	Prepare only freshly slaughtered. Proceed as for joints of beef.	As for joints of beef.
Chops	As for lamb chops.	As for lamb chops.
Veal Joints	As for joints of beef.	As for joints of beef.
Escalopes	Flatten, egg and crumb. Separate with foil or moisture-vapour-proof paper dividers. Wrap tightly.	Moisture-vapour-proof paper or freezer foil. Over-wrap in freezer bag. Seal with tie.

Place the foil on top of the tray, making sure the tray is sitting right in the centre and foil extends well over all the sides.

Now twist the foil over and over working right round the tray, lodging the foil under the lip-edge to ensure a perfect seal.

MEAT Offal	PREPARATION	PACKAGE
Heart Kidney Liver Sweet- breads	Soak in cold water for 30 minutes. Drain and rinse. Remove fat, valves and other tissues. Divide into suitable portions. Place in freezer bags.	Freezer bags. Seal with ties or by heat.
Tripe	Soak in cold water for 30 minutes. Cut into 1-inch cubes. Divide into suitable portions.	As above.
Sausages	Wrap tightly. Seal and freeze. Store for short period only.	Sheet polythene or freezer bags. Seal with ties or by heat.
Sausage meat	Omit seasoning. Divide into suitable portions. Wrap closely.	Moisture-vapour-proof paper. Seal with freezer tape.
Poultry Chicken, whole	Pluck and clean. Wipe inside with clean cloth. Cover protruding bones with foil. Wrap giblets separately and place in body cavity.	Pack in freezer bag. Extract as much air as possible. Seal with tie or by heat.
Chicken pieces	Egg and crumb or leave plain.	Plastic containers.
Duck, whole	As for whole chicken.	As for whole chicken.
Duck pieces	As for chicken pieces.	As for chicken pieces.
Game Pheasant Grouse	Hang for the required amount of time. Pluck and clean. Proceed as for whole chicken.	As for whole chicken.
Hare Rabbit	Bleed and hang as required. Skin, clean and joint. Pack individual joints or in suitable portions.	Pack individually in polythene and over-wrap with stockinette or pack into plastic containers.

while it is thawing, and this must be taken into account.

It is advisable, though not essential, to pluck or skin and draw game before freezing, as handling afterwards is a none too pleasant a job.

Individual instructions for freezing game

Rabbits and hares

Behead and bleed, then hang in a cool place for 24 hours. After hanging, skin and disembowel, then wash out the interior and drain. Wipe the outside with a damp cloth, and joint. Wrap each joint separately, eliminating pockets of air, then put all the joints together in one large container or suitable bag; seal, label and freeze.

Game birds

Hang the bird for one day less than usual, then pluck and take out the shot. Draw and wash the interior. Drain. Wipe the outside with a damp cloth, then pack in a suitable material like cling wrap. Seal, label and then chill before freezing.

VEGETABLES

Certain basic principles apply in freezing both vegetables and fruits. This concerns the need to plan well ahead, for example, for there are vegetable gluts that must be allowed for, if you want to make sure you have space in the freezer when they come along. For instance, if you freeze Brussels sprouts at the peak of their perfection and when the prices are low, you can have them, full of colour and flavour, at any time of the year. As with the fruit gluts, make sure that you use up all one year's supply before the next season comes round.

Your choice of vegetables must be guided first and foremost, by their quality. Bruised or marked vegetables should be discarded as must those that are not as fresh as they could be. Speedy freezing is still of prime importance and the sooner after picking that the vegetables are in the freezer, the better.

Here is a tip that will save you time and possible disappointment. Sort through the vegetables chosen for freezing, put unsatisfactory specimens on one side. Prepare these separately for immediate use. If there are too many prepared for one meal, washed and strained vegetables can be stored in polythene bags or containers in the refrigerator for at least a week, and used as required. A local farmer or grower may be willing to sell you bulk supplies of vegetables during their glut if you are unable to grow your own.

General principles in preparation

In the main, vegetables that are usually cooked before eating freeze well. Vegetables that are usually eaten raw for their crisp texture and delicate flavour, or which hold a lot of water, are not suitable for freezing. Into this category come celery, cucumber, lettuce, radishes, spring onions, and large marrows. Baby marrows may be frozen successfully. If you have some doubt as to whether your family will or will not like a certain untried vegetable, do test it on them before freezing large quantities! Frozen vegetables are simplicity itself to thaw, as all the preparation work is done before freezing take place.

If you have too many vegetables to prepare at one time you can store them for a short period in the refrigerator – just until you have the first batch out of the way and are free to prepare the remainder. Speed from garden to freezer is always important and short storage in the refrigerator is only recommended to help you out of a difficulty. Prepare all the vegetables straight away, if you possibly can. The same applies to ready-prepared vegetables which the freezer is unable to freeze at once. Store these in the refrigerator, until the freezer is able to take more unfrozen packets.

Pack the vegetables in quantities suitable for use on one occasion so that you have no leftovers. Tupperware 16-oz. and 30-oz. square rounds are useful here. Vegetables cannot be re-frozen once they are thawed so if you have any thawed ones left over, put them in the refrigerator and eat them as soon as you can, certainly within 2–3 days.

Pack several vegetables together so that you have a few packs of mixed vegetables such as diced carrots, peas, chopped beans.

Alternatively, freeze the vegetables by the loose pack method. This is done by spreading out the prepared cooled vegetables on baking sheets or large shallow pans and placing them in the freezer, without covering and sealing. When frozen (about 1–2 hours) scrape the vegetables loose from the trays and pour into polythene bags, seal, leaving no headspace, label and return at once to the freezer before they can begin to thaw. A pack may then be re-opened, a small amount removed and the remainder re-sealed and returned to the freezer.

Because the vegetables are already part-cooked in the preparation process, remember cooking time is reduced by half to one third.

There are several methods of cooking frozen vegetables – cooking gently in a little boiling water, steaming, baking, sautéing and pressure cooking. These methods have been described fully in Chapter Six, in a section beginning on page 73, together with cooking time-tables for both the boiling water and pressure cooking methods.

Packing in polythene bags

When polythene bags are used, it is necessary to extract as much air as possible before sealing. A simple way to do this is to draw the bag up to form a 'neck', and wind with a wire fastener, loose enough to insert a drinking straw into the bag. Suck out as much air as possible through the straw, then quickly tighten the wire fastener to prevent air from returning to the bag. Label in the usual way and freeze. The step-by-step photographs on preparing chips for the freezer on page 49, show this being done. Alternatively freeze in special bags that can be taken from the freezer and dropped into boiling water to complete cooking.

Varieties most suitable for freezing

Broad beans

Most varieties react satisfactorily to freezing as long as they are not over mature.

Stringless and Runner beans

TENDERGREEN and TENDERLONG are considered to be the best stringless beans. EMERGO, EMPEROR, CRUSADER and NE PLUS ULTRA are the best runner beans.

Carrots

EARLY NANTES is considered the best with PERFECT GEM second and SCARLET INTERMEDIATE or any short horn varieties when young and tender.

Corn on the cob

JOHN INNES HYBRID or any quick maturing variety.

Peas

EARLY ONWARD, MIRACLE, VICTORY FREEZER, KELVEDON WONDER.

Spinach

GOLIATH, NEW ZEALAND or any perpetual spinach react well.

Brussels sprouts

The sprouts should be small and firm. The following varieties give good results: CAMBRIDGE No. 3, CAMBRIDGE SPECIAL, IRISH ELEGANCE, JADE CROSS.

The following chart indicates the approximate storage space required for 1 lb. vegetables.

Type of vegetables	Capacity of container
Asparagus	$1\frac{3}{4}$ pints
Aubergines	2 pints
Beans, broad	$1\frac{1}{4}$ pints
Beans, runner	1 pint
Broccoli	2 pints
Brussels sprouts	$1\frac{1}{2}$ pints
Cauliflower sprigs	2 pints
Peas	1 pint
Spinach	$\frac{1}{4}$ pint

Blanch cauliflower flowerets and dry on kitchen paper.

Pack flowerets in 30-oz. square round containers.

How to blanch

Blanching is the most important step in preparing vegetables for the freezer, for this process is responsible for slowing down the action of the enzymes which eventually cause deterioration. As an extra bonus, blanching also softens the vegetables so that they become easier to pack.

To blanch in boiling water, use a large saucepan and a wire mesh basket (preferably a rigid or collapsible blancher) to hold the vegetables, or some stockinette. You need 1 gallon of fast boiling water for each pound of prepared vegetables. Preferably blanch only a pound at a time to ensure it is done properly. Put the vegetables in the wire basket or stockinette and immerse in the fast boiling water. Keeping the heat high, blanch for the exact time given for each vegetable. Timing begins the instant the water returns to the boil. It must be accurate as underblanching will not destroy the enzyme activity and overblanching may spoil the flavour and texture of the vegetable. Be sure that the water returns to boiling before blanching more vegetables and change the water after 6–8 blanchings.

Cool quickly and thoroughly.

An alternative method is to put the vegetables straight into a pan of boiling water and blanch, strain into a colander and cool.

To blanch in steam place a thin layer of vegetables in the mesh basket or stockinette. Fill a large saucepan to a depth of 1 inch; place a rack or trivet on the bottom so that the vegetable basket or stockinette will rest on this, above the water level. Bring the water to a fast boil and put the vegetables on the trivet; cover with a lid and time the blanching time accurately. Steam blanching takes half as long again as the boiling water method, so that a vegetable recommended for 2 minutes' blanching in boiling water, needs 3 minutes if it is to be steam blanched. Don't steam blanch leafy vegetables like cabbage and spinach, because they tend to mat together. Cool quickly.

You needn't always blanch

Certain vegetables can be stored for a *short* time without blanching – French beans, carrots, courgettes, mushrooms, onions, peas, peppers, spinach, tomatoes.

If you don't blanch, remember that these vegetables will need their full cooking time when they are being prepared for the table. Unblanched vegetables still contain certain micro-organisms and so they must be cooked quickly after thawing or spoilage will occur.

How to cool blanched vegetables

Place them under a stream of cold water or in a pan of cold water, immediately after taking them from the blanching pan. The quickest way of cooling the vegetables is to place them under cold running water and then put the cooled, drained vegetables in a container and surround it with ice-cubes. Not everyone has the facilities for making sufficient cubes to do this but, if it is pos-

sible, it ensures speedy chilling, a vital factor in good frozen vegetables.

Blanching time-table Note: When no time is given for steam blanching it is felt that the vegetable is unsuitable for this method.

Vegetable	Boiling water Minutes	Steaming Minutes
Asparagus, small spears	2	3
Asparagus, medium spears	3	4½
Asparagus, large spears	4	6
Artichokes (According to size)	5–7	–
Aubergine	4	–
Beans, Broad	3	4½
Beans, French or Runner		
whole	2–3	3–4
sliced	1	2
cut	2	3
Broccoli, thin	3	4
Broccoli, medium	4	5
Broccoli, thick	5	6
Brussels sprouts, small	3	–
Brussels sprouts, medium	4	–
Cabbage (sliced)	1½	–
Carrots (sliced or diced)	3	–
Carrots (whole)	5	–
Cauliflower	3	5
Celery	3	–
Corn on the cob (according to size)	5–8	–
Parsnip	2	–
Peas	1–1½	1½–2
Spinach	2	–
Turnips	2½	4

Wet packing or dry?

Although dry packing is the most commonly used and recommended, some experts feel that wet packing in brine makes the vegetables less likely to toughen while frozen. It is therefore entirely a matter of personal choice if you wish to try packing in brine. Start with just a small amount so that if the results are not to your liking, you haven't a freezer brim full of brine packed vegetables to eat up! Prepare the vegetables in the usual way and, after they have been blanched, cooled and drained, pack them in polythene containers, leaving $\frac{1}{2}$–1-inch headspace. Cover with cold brine, made up from 2 tablespoons of salt to 1 quart of water. Seal, label and freeze. In the instructions that follow, the dry method of packing is used. To brine-pack, simply add the solution at the appropriate stage.

Individual instructions for vegetables

Asparagus

Select young, tender stalks with compact tips, wash thoroughly then trim off tough part of stalk. Blanch, cool, drain and package the spears alternating tips and stem ends. Leave no headspace, seal, label and freeze.

Artichokes

Remove all the green leaves and the centre flower, then wash thoroughly. Blanch in boiling water with 1 tablespoon lemon juice added. Cool, pack the bases in polythene bags or rigid polythene containers, seal, label and freeze. Use the leaves for soup.

Aubergine

Peel and cut into 1-inch slices. Blanch, chill, and pack into rigid polythene containers. Leave a $\frac{1}{2}$-inch headspace. Seal, label, freeze.

Beans, Broad

Pod and pick over the beans carefully removing those unsuitable for freezing. Blanch, drain and pack in polythene bags. Seal with wire fasteners, label and freeze. If a rigid container is used, leave a $\frac{1}{2}$-inch headspace.

Beans, French or Runner

Select tender, young beans that are not stringy. Top and tail and wash thoroughly in cold water. The beans may be frozen cut, sliced or whole. The blanching time depends on the method you have chosen. After blanching, cool quickly and drain, then pack leaving a $\frac{1}{2}$-inch headspace. Seal, label and freeze.

Beetroot, young

Wash the beetroot and trim, leaving about $\frac{1}{2}$ inch of the stem. Cook in boiling water until tender. The time this takes naturally depends on the size of the beetroot, but for small ones allow about 25–30 minutes while the slightly larger ones will take 45–50 minutes. Freezing beetroots that are more than 3 inches across is not recommended. Care should be taken that they don't bleed. Cool quickly, peel and cut into dice or slices. Pack leaving $\frac{1}{2}$-inch headspace, seal, label and freeze.

Broccoli

Best for freezing is broccoli with tight, compact heads, dark green in colour, tender stalks with no hint of woodiness. Wash, peel the stalks and trim. Make up a salt and water solution (4 teaspoons to 1 gallon cold water) and soak the broccoli in this for half an hour to remove any insects. Split each piece down the centre into a more manageable size, then blanch. Cool quickly, drain and pack without leaving a headspace. Seal, label and freeze.

Brussels sprouts

Choose small sprouts with firm compact heads and a good green colour. Remove discoloured leaves and wash. Sort out into sizes, making sure you freeze even-sized sprouts in the same pack. Blanch, cool and drain. Pack leaving no headspace, seal, label and freeze.

Cabbage

Frozen cabbage is not suitable as a salad vegetable, say in the form of coleslaw. To be successful it must be used cooked. Choose firm, solid cabbage and trim off the outer leaves. It can be frozen shredded, cut into wedges or with the leaves separated. Blanch, cool, pack leaving $\frac{1}{2}$-inch headspace. Seal, label and freeze.

Carrots

Young, tender carrots with a mild flavour are best for freezing. Trim tops, wash thoroughly then peel, or rub off skins. The very small carrots may be left whole, the larger ones (though the really large carrots should not be frozen at all) may be diced or sliced. Blanch, cool, pack leaving ½-inch headspace, seal, label and freeze.

Cauliflower

Select firm, tender cauliflower with good coloured heads. Trim, then break the cauliflower into small flowerets about an inch across. Soak for half an hour in salt and water solution (4 teaspoons salt to 1 gallon water) to remove the insects, then wash well and drain. Blanch in a salt and water solution of the same proportions used for soaking, cool and drain. Pack, leaving no headspace, seal, label and freeze.

Celery

This vegetable is not suitable for freezing if it is to be used as a salad vegetable or eaten raw in any other way. You can freeze it if it is to be cooked before eating. Choose crisp and tender celery and make sure it isn't stringy. Wash

carefully to remove all the grit, trim and cut into 1-inch pieces. Blanch, cool, drain and pack, leaving a ½-inch headspace. Seal, label and freeze.

Corn on the cob

Choose corn which is plump and tender. Trim off the ears and silk then wash. Blanching time varies according to the size of the cob so that they must be graded for size. Blanch together cobs of similar size – small cobs 5 minutes, medium ones 6½ minutes, and large ones 8 minutes. After blanching, cool quickly and drain. Wrap in a cling wrap material. Seal, label and freeze.

Mushrooms

Select fresh mushrooms as near perfect as possible, certainly without any decay. Wash them carefully in cold water or merely wipe them and trim off the ends of the stems. Small button mushrooms can be frozen whole, but larger ones (those more than 1 inch across) should be quartered or sliced. To preserve the colour, dip the mushrooms in a lemon and water solution (1 teaspoon lemon juice or 1½ teaspoons citric acid to 1 pint of water) for 5 minutes. Then steam blanch – 5 minutes for whole mushrooms, 3½ minutes for mushroom quarters, 3½ minutes also for button mushrooms, 3 minutes for slices. Cool and drain. Pack, seal, label and freeze.

There are two other methods of preparation. For the first, wash, trim and slice the mushrooms, if necessary. Don't add to lemon and water solution, but heat the mushrooms in butter in a frying pan until almost cooked. Heat small

Strain blanched peas in a colander under running cold water. Shake dry.

Spoon peas into shallow 16-oz. square round containers for family-sized portions.

For colour identification pack peas in stripy polythene bags with pads of folded foil to give a semi-rigid base. For large bags place foil divider half way up the pack.

Country house chicken (page 104)

Pear condé and Gingered pear coupes (page 114)

Peel, wash and chip the potatoes. Part-fry them in deep fat or oil, until just soft but not browned.

Drain and cool the chips, then spoon them into a plastic bag.

Loosely tie the bag with a wire fastener, leaving just enough room for a drinking straw to be inserted.

Draw all of the air out of the bag through the drinking straw, then quickly tighten the wire fastener.

quantities at a time. For the other method, trim and blanch for 30 seconds, drain and cool quickly. Pack, leaving a $\frac{1}{2}$-inch headspace, seal, label and freeze.

Parsnips

Tender young parsnips are best for freezing and choose those that are free from woodiness. Trim the tops, wash, then peel thinly. Slice, blanch, cool and drain. Pack, leaving a $\frac{1}{2}$-inch headspace, seal, label and freeze.

Peas

Shell. Discard the peas that are old or large, freeze only those that are tender and sweet. Blanch, cool and drain. Pack, leaving a $\frac{1}{2}$-inch headspace. Seal, label and freeze.

Peppers

If the peppers are to be used uncooked, they should be packed without blanching. Simply wash, remove stems, halve, and take out the seeds. Slice, label and freeze. For peppers that are eventually to be cooked, prepare in the way described here, but blanch and cool them before packing. Blanched peppers must have a $\frac{1}{2}$-inch headspace. Seal, label and freeze.

Potatoes

For chipped potatoes wash, peel, and slice in the usual way. Then deep fry them until soft but not browned. Drain really well, then cool. Pack into polythene bags using the drinking straw method for extracting all the air out of the bag before sealing. Seal, label and freeze.

Tiny boiled new potatoes can be packed by the same method. Old potatoes should be cooked, mashed with milk and beaten egg then piped into rosettes on a baking tray. Open freeze and pack in rigid based containers. Roast potatoes should be removed from the oven light brown, and frozen in bags. Thaw and reheat in a roasting tin in a hot oven for 25–30 minutes.

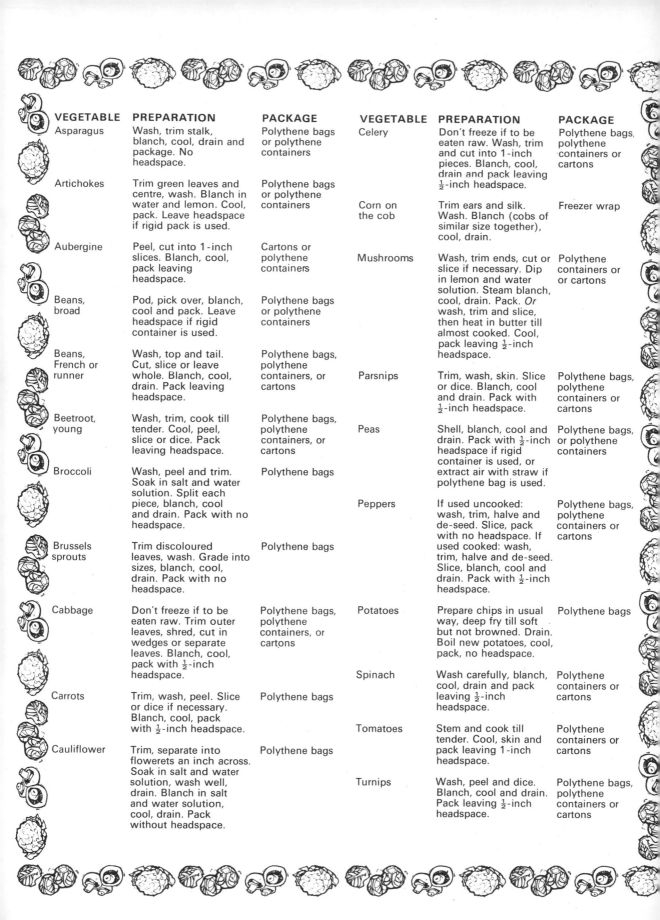

VEGETABLE	PREPARATION	PACKAGE	VEGETABLE	PREPARATION	PACKAGE
Asparagus	Wash, trim stalk, blanch, cool, drain and package. No headspace.	Polythene bags or polythene containers	Celery	Don't freeze if to be eaten raw. Wash, trim and cut into 1-inch pieces. Blanch, cool, drain and pack leaving $\frac{1}{2}$-inch headspace.	Polythene bags, polythene containers or cartons
Artichokes	Trim green leaves and centre, wash. Blanch in water and lemon. Cool, pack. Leave headspace if rigid pack is used.	Polythene bags or polythene containers	Corn on the cob	Trim ears and silk. Wash. Blanch (cobs of similar size together), cool, drain.	Freezer wrap
Aubergine	Peel, cut into 1-inch slices. Blanch, cool, pack leaving headspace.	Cartons or polythene containers	Mushrooms	Wash, trim ends, cut or slice if necessary. Dip in lemon and water solution. Steam blanch, cool, drain. Pack. *Or* wash, trim and slice, then heat in butter till almost cooked. Cool, pack leaving $\frac{1}{2}$-inch headspace.	Polythene containers or or cartons
Beans, broad	Pod, pick over, blanch, cool and pack. Leave headspace if rigid container is used.	Polythene bags or polythene containers			
Beans, French or runner	Wash, top and tail. Cut, slice or leave whole. Blanch, cool, drain. Pack leaving headspace.	Polythene bags, polythene containers, or cartons	Parsnips	Trim, wash, skin. Slice or dice. Blanch, cool and drain. Pack with $\frac{1}{2}$-inch headspace.	Polythene bags, polythene containers or cartons
Beetroot, young	Wash, trim, cook till tender. Cool, peel, slice or dice. Pack leaving headspace.	Polythene bags, polythene containers, or cartons	Peas	Shell, blanch, cool and drain. Pack with $\frac{1}{2}$-inch headspace if rigid container is used, or extract air with straw if polythene bag is used.	Polythene bags, or polythene containers
Broccoli	Wash, peel and trim. Soak in salt and water solution. Split each piece, blanch, cool and drain. Pack with no headspace.	Polythene bags	Peppers	If used uncooked: wash, trim, halve and de-seed. Slice, pack with no headspace. If used cooked: wash, trim, halve and de-seed. Slice, blanch, cool and drain. Pack with $\frac{1}{2}$-inch headspace.	Polythene bags, polythene containers or cartons
Brussels sprouts	Trim discoloured leaves, wash. Grade into sizes, blanch, cool, drain. Pack with no headspace.	Polythene bags			
Cabbage	Don't freeze if to be eaten raw. Trim outer leaves, shred, cut in wedges or separate leaves. Blanch, cool, pack with $\frac{1}{2}$-inch headspace.	Polythene bags, polythene containers, or cartons	Potatoes	Prepare chips in usual way, deep fry till soft but not browned. Drain. Boil new potatoes, cool, pack, no headspace.	Polythene bags
Carrots	Trim, wash, peel. Slice or dice if necessary. Blanch, cool, pack with $\frac{1}{2}$-inch headspace.	Polythene bags	Spinach	Wash carefully, blanch, cool, drain and pack leaving $\frac{1}{2}$-inch headspace.	Polythene containers or cartons
Cauliflower	Trim, separate into flowerets an inch across. Soak in salt and water solution, wash well, drain. Blanch in salt and water solution, cool, drain. Pack without headspace.	Polythene bags	Tomatoes	Stem and cook till tender. Cool, skin and pack leaving 1-inch headspace.	Polythene containers or cartons
			Turnips	Wash, peel and dice. Blanch, cool and drain. Pack leaving $\frac{1}{2}$-inch headspace.	Polythene bags, polythene containers or cartons

Spinach

Fresh tender spinach should be chosen and special attention must be paid to washing it. Trim off the coarse stalks. Blanch a small amount at a time and occasionally shake the blanching basket to make sure the heat penetrates every leaf and keeps them separated. Cool, drain and pack leaving a ½-inch headspace. Seal, label and freeze.

Tomatoes

Choose ripe but firm tomatoes, remove stems. Peel and quarter. Put the quarters in a pan, cover and cook for about 15 minutes until tender. Put the pan in cold water to allow the tomatoes to cool, then pack them leaving approximately 1-inch headspace (more if a container larger than a pint is used). Seal, label and freeze. For freezing tomato juice, see the section for fruit juices on page 59.

Turnips

Turnips with a mild flavour are better for freezing and you should select young and tender ones that are no bigger than small to medium. Wash them, then peel and dice. Blanch, cool and drain. Pack, leaving ½-inch headspace, seal, label and freeze.

DAIRY FOODS
Eggs

It is not possible to freeze eggs if left in their shells because the shells will crack. Nor is it satisfactory to freeze hard-boiled eggs because the white tends to become rubbery and unpalatable. You can, however, successfully freeze raw shelled eggs, whole or separated.

First, choose fresh eggs and wash carefully in cold water, since dirt remaining on the shell could cause contamination. Cracked eggs must be discarded because bacteria could already have been introduced into the egg through the crack. Before adding the eggs to a large bowl, crack each one into a cup so that any bad eggs may be thrown away. If a bad one was added to the large bowl, it would taint the remainder and they would all have to be discarded.

Whole eggs

Shell eggs and put into a bowl. Beat, but do not whip in air. Add salt or sugar to prevent the mixture thickening – 1 teaspoon salt to ¾ pint of eggs to be used for savoury dishes, 2 tablespoons sugar to 1 pint of eggs, for desserts or baking. Don't forget to note on the packet whether it is sweet or savoury egg. Once the eggs are beaten they are ideal for omelettes, scrambled eggs, cakes and puddings.

Egg whites

Simply separate the whites from the yolks and freeze them in suitable containers without beating first or adding salt or sugar. They are excellent for making meringues when used at room temperature.

Egg yolks

Separate the yolks from the whites and beat them without whipping in air. Add salt or sugar, depending on later use, in the proportion of 1 tablespoon sugar or ½ teaspoon salt to each cupful of yolk. Note which you have done on the outer wrapping.

Packaging eggs

Pack the eggs in containers that hold just the amount you will want to use at any one time. Once thawed, the eggs cannot be re-frozen. Allow a little headspace in the container for expansion. A good way of freezing eggs into handy quantities is to pour the mixture into a plastic ice-cube tray, freeze, then remove the egg cubes and store in polythene bags; this makes it easy to extract the number you need and return the rest to the freezer. Alternatively, small plastic egg cups may be used in the same way.

Milk

Although it is possible to freeze milk it is not always entirely satisfactory. Our usual day-to-day milk is liable to suffer great loss of quality because the fat tends to separate at freezing temperatures. Sometimes homogenised milk seems to freeze quite well for a short period and it is worth trying to see how successful you are. The only type of milk that can really be recommended for freezing without loss of quality is pasteurised homogenised milk. Freeze it in suitable cartons leaving a 2-inch headspace for expansion. This cannot be regarded as a long-term method of keeping milk – it should be used up in 2 weeks.

Cream

Like some milk, low butterfat cream tends to separate when frozen. However, double cream (over 40% butterfat) may be frozen satisfactorily. This also applies to clotted creams as they have already been pasteurised. Once frozen, cream may be stored for up to 4 months if a little sugar is added to increase the keeping time. (1 tablespoon sugar to 1 pint of cream.) Store in suitable containers allowing 1-inch headspace for expansion. The cream should be frozen at the earliest possible moment, and the actual preparation should be completed before any deterioration can take place. It is possible to freeze the cream from the top of the milk, but once thawed it must be beaten vigorously and served as a whipped cream accompaniment.

Butter

You may freeze commercially sold butter in its original wrapping if it is still firm, but if it has become soft, overwrap with a suitable freezer material such as cling wrap. Home-made butter should be packed in cartons or special freezer wrapping. Storage time varies according to the type of butter to be frozen. Unsalted butter made from pasteurised cream can be stored for up to a year. Salted butter will keep for only 6 months.

Cheese

Not all cheeses are suitable for freezing, especially the soft cream cheeses, though if they are formed into a dip by blending with other foods they can be stored satisfactorily for up to 2 months. The most satisfactory ones to freeze are the hard cheeses like Cheddar, which should be divided up into half-pound portions before wrapping, so that you don't thaw too much in one go. They can be kept for 4–6 months. Several of the less common cheeses freeze well. These include Gruyère, Port du Salut, Mozzarella, Stilton, Emmenthal, Parmesan, Gouda, Derby, Roquefort, Edam. Blue cheeses freeze well, but are inclined to become crumbly when thawed. If this happens, use for salads or sprinkled cheese toppings for other dishes. So that cheese is at its very peak once thawed, allow the cheese to mature until it is just right for your palate before freezing. Alternatively, buy mature cheeses in the first place. It is most important to pack cheese carefully to avoid risk of cross-contamination and drying out.

Ice cream

Many freezer owners, especially those with children, buy a big container of ice cream to keep on hand – a wonderful convenience for those tempting summertime desserts. Ice cream can now be bought in many exciting flavours as well as vanilla; mouthwatering rum and raisin, the smooth luxury of honey and brandy, and many more. They come in plastic containers with seals in a variety of sizes, ½-litre, 1-litre and 5-litre. It is useful to have various sizes of packs available and keep some in the frozen food storage compartment of the refrigerator for quick use, as it will be less solidly frozen – it isn't too easy attacking a large and very solid block of ice cream for a couple of servings! Press a piece of polythene tightly against the remaining ice cream to protect it from air.

Return the container to the freezer quickly before thawing begins. Don't re-freeze ice cream once it has melted.

Home-made ice cream may also be frozen.

Leftover wine

Small quantities should be packed in rigid based containers with a headspace and carefully labelled, as sweet wines are not suitable for savoury dishes. De-frost before adding to the dish or turn the wine cube straight into pan.

FRUIT

For seasonal foods like fruit, you need to plan well ahead to cope with the various gluts while, at the same time, making use of all the freezer space available. For example, make sure you have space in the freezer if you want to take advantage of the yearly glut of strawberries. Use up all the previous year's supply before the next season is upon you. Good management of the freezer requires experience and thought, but if you can estimate how much of any one fruit you are to freeze, and how often you are likely to serve it, you can work out a plan of campaign. To give you some idea of how this works – thirty family-size packages of any one fruit should be enough to serve that family once a week from October to May.

Like every other food destined for the freezer, fruit must be carefully selected and prepared for freezing. Most fruits are suitable, but those with pronounced flavours freeze best of all – blackberries especially, peaches, strawberries and rhubarb.

Although people without their own gardens may not have really fresh fruit immediately on hand, the 'not more than two hours from garden to freezer' rule is still the best. It may be possible to buy good quality fruit from a farmer if you offer to pick it yourself.

Varieties most suitable for freezing

Strawberries
CAMBRIDGE VIGOUR followed by ROYAL SOVEREIGN but it must be remembered that all varieties give better results if frozen sliced or puréed.

Strawberries being packed in cooled sugar syrup in a 30-oz. Tupperware square round. The container is filled to allow ½-inch headspace.

Raspberries
NORFOLK GIANT giving best colour and flavour, with LLOYD GEORGE a close second.

Plums
VICTORIA plums and the purple varieties are considered to be far the best, both raw or cooked after freezing.

Apples
BRAMLEY SEEDLING is considered the best variety for freezing.

Pears
Pears do not on the whole freeze well. When thawed they tend to be either too hard or too soft.

Greengages
CAMBRIDGE GAGE gives a good flavour, but is variable in colour. COUNT ALTHANN'S GAGE gives good colour, but weaker flavour. JEFFERSON gives good colour and flavour, but the skins tend to be tough.

The following chart indicates the approximate storage space required for 1 lb. fruit.

Type of fruit	Container space required
Apples, sliced	2 pints
Blackberries	1¼ pints
Cherries	1½ pints
Gooseberries	1½ pints
Loganberries	1¼ pints
Peaches (sliced)	1½ pints
Pineapples (sliced)	2 pints
Raspberries	1¼ pints
Rhubarb	1½ pints
Strawberries	1½ pints

Dry packs
Unsweetened or dry packs are only successful when fruit can be prepared without breaking the skin, or if the fruit does not lose its colour while you are preparing it. It simply means that the fruit is packed carefully into waxed boxes, cartons or bags (leaving the usual ½-inch headspace for expansion), sealed and placed directly into the freezer.

Sugar packs

For soft fruit with plenty of juice, a sugar pack is the best method. First pick over, wash and drain the fruit, if necessary, then put a little in the carton or box. Follow this with a layer of sugar, then another layer of fruit and so on until all the fruit and sugar is used up (in the proportion of 3–5 lb. of fruit to 1 lb. of sugar). Now cover the box and seal with special freezer tape. Label the package before putting it into the freezer. Handy Tupperware containers in small round and square round shapes are ideal for packing berry fruits. An alternative method is to place all the washed and drained fruit with the sugar in one large bowl; turn the fruit gently with a wooden spoon until each piece is coated in sugar. Then transfer to the container in which you intend to freeze it.

Syrup packs

Some fruits discolour very easily during preparation and these are the ones that call for a syrup pack – as do the fruits with little juice. The syrup is prepared in varying strengths, according to the type of fruit to be frozen and individual preference.

A heavy syrup, for example, known as a 50% syrup, is made with 1 lb. of sugar to 1 pint of water. A medium syrup, and incidentally the one most commonly used, is a 40% syrup made with 11 oz. sugar to 1 pint of water. Occasionally, a weaker syrup of some 20–30% is required, either because you are freezing fruit which is much more delicately flavoured, or simply because you prefer it that way.

Dissolve the sugar in hot or cold water and remember to chill it before pouring on to the fruit. The easiest way to make quite sure that the syrup is sufficiently chilled is to make it up the day before you need it and keep it in the refrigerator ready for use.

The following table gives you the sugar and water proportions that are needed for different strengths of syrup.

For different strengths of syrup use:

Solution	Sugar	Water	Strength
10%	2 oz.	1 pint	Very thin
20%	4 oz.	1 pint	Thin
30%	7 oz.	1 pint	Medium thin
40%	11 oz.	1 pint	Medium heavy
50%	16 oz.	1 pint	Heavy
60%	25 oz.	1 pint	Very heavy

Use $\frac{1}{4}$ pint of syrup to each 1-pint package of fruit. The fruit should be sliced directly into the cold syrup and be completely covered, for if any pieces remain uncovered they may discolour or lose flavour. Place a piece of crumpled foil on top of the fruit to keep it under the syrup, then cover, seal and label.

Preparing fruit for the freezer

The fruit you select for freezing must be fully ripe but still firm. First, wash it in cold water and handle delicate fruits such as blackberries in small quantities to avoid bruising or squashing. Remove the fruit from the water and drain thoroughly. Less bulky packs can be made up where it is possible to slice the fruit first. There is the added bonus of finding that the fruit is ready to use, once thawed. Slicing also means that you can freeze fruit that is only partly suitable. Simply cut away the bruised area, then slice and freeze the remainder. Don't feel that you can never use fruit that is a little over-ripe for this may still be frozen if it is first crushed or puréed.

Try to prepare only as much fruit as your freezer can freeze in one go. But if you find that you have some which cannot be put straight into the freezer, pack it and preserve in the refrigerator only until the freezer will take more unfrozen packages. A quart container will take approximately $1\frac{1}{2}$–$1\frac{3}{4}$ lb. of fruit so if you know your freezer will freeze a certain number of containers you can work out easily how much fruit should be prepared.

Don't use galvanised, iron or copper utensils to prepare fruit because these metals taint flavours and cause discoloration. The same applies to enamel utensils that are chipped or tin ones that are not well tinned.

Preventing discoloration

Once the preparation process begins, the cut surfaces of light coloured fruits, such as peaches and apples, turn dark very quickly. Ascorbic acid ($\frac{1}{4}$ teaspoon to $\frac{1}{2}$ pint cold water) is one of the best preventatives you can use against discoloration since it preserves natural colour and flavour, and adds nutritive value as well. Ascorbic acid can easily be bought, by the ounce, from most chemists.

Lemon juice is another possible colour preserver. It contains both ascorbic acid and citric acid but to be effective a lot of lemon juice is needed, and this will give the fruit a sour flavour which may be objectionable to some people.

Apple slices can be steamed for a few minutes before packaging to prevent browning, and some fruits can be blanched in boiling syrup for 1–3 minutes. The heat inactivates enzymes that cause discoloration and wilting.

Individual instructions for fruit

Apples

Wash, pare, core and slice the apples $\frac{1}{2}$ inch thick. Put into salted water, or blanch the slices in boiling water for 1 minute (slightly longer if the apples are very firm). Do not pulp by over-cooking. Yet another alternative is to place the apple slices in a single layer in a steamer and steam from $1\frac{1}{2}$–3 minutes depending on thickness of the slice. Sprinkle 6 oz. sugar evenly over each quart of apples. Press apples into containers leaving headspace. Seal, label and freeze. An unsweetened pack can be frozen by the same method – omitting the sugar.

Alternatively, slice the apples straight into a container quarter-filled with a cold 40% syrup. Cover with crumpled foil to keep fruit under the syrup, leave headspace, seal, label and freeze.

Apricots

Select ripe, but firm apricots that are evenly coloured. Wipe, halve and stone. Apricots are subject to discoloration and an ascorbic acid solution of $\frac{1}{4}$ teaspoon ascorbic acid to $\frac{1}{2}$ pint of cold water should be poured over each 2 lb. fruit.

Small quantities of table wine left in the bottle can be frozen for future use in cooking. Choose wide-necked containers, allow headspace and label whether sweet or dry.

A mousse made with egg yolks and sugar can be flavoured with wine or fruit purée and frozen in pretty glasses covered with foil caps or sealed in Tupperware dessert dishes.

Remove caps or seals while the mousses are still frozen, and pipe with cream. Add pretty finishing touches and serve slightly chilled.

For a dry sugar pack, stir in 4 oz. of sugar to each pound of fruit, or layer the fruit and sugar in the container itself. Leave headspace, seal, label and freeze. If a syrup pack is preferred, put the fruit into containers and cover with a 40% syrup. Place crumpled foil on top to keep fruit under the syrup, leave headspace, seal, label and freeze.

Avocados

These discolour when frozen whole. Pulp the flesh, mix with lemon juice or chicken stock for cold soups. Seal in containers, label and freeze.

Blackberries

Select firm, plump, fully ripe, but not over-ripe berries with glossy skins. Green berries may spoil the rest of the fruit. Wash and drain the fruit and, if to be eaten uncooked, pack into container and cover with cold 40–50% syrup depending on sweetness of the fruit. Leave headspace, but add crumpled foil to hold fruit under syrup. Seal, label and freeze. If berries are to be used later for jam, freeze in a dry sugar pack in the proportion of 1 quart fruit to 6 oz. sugar. Turn berries over and over in the sugar until most of it is dissolved. Fill containers, leaving headspace. Seal, label and freeze.

Cherries

Choose tree-ripened cherries with a good red colour (red cherries are better suited to freezing than black ones). Remove stalks, wash, drain and remove stones. Pack into containers and cover with cold 40–50% syrup depending on sweetness. Leave headspace. Seal, label and freeze. Alternatively, freeze in a dry sugar pack using 4–6 oz. sugar to 1 quart cherries. Turn fruit over in the sugar until it is dissolved and pack into containers. Leave headspace, seal, label and freeze.

Cranberries

Remove the stalks, take out the berries that are less than perfect, wash the remainder and drain. Pack into containers without sugar, seal, label and freeze.

Currants

Both blackcurrants and redcurrants are treated in the same way. Top and tail, wash and drain. Pack into containers without sugar, seal, label and freeze. For a dry sugar pack, mix fruit and sugar (in the proportion of 3 parts currants to 1 part sugar) until sugar is almost dissolved, then pack into containers leaving headspace, seal, label and freeze. Currants may also be stored in a 40–50% syrup. Simply pack into containers then pour the cold syrup over until the fruit is covered. Add a piece of crumpled foil before covering so that the fruit is kept under the syrup, seal, label and freeze.

Damsons

This fruit is not suitable for freezing in the normal way because the skins are liable to become tough and the stone may well flavour the fruit. Unless they are packed as a purée, damsons can only be stored, unsweetened, for a short time.

Gooseberries

Top and tail, wash and pack into a dry sugar pack. Do this by putting a layer of gooseberries, then a layer of sugar, then more gooseberries and so on, finishing with a layer of sugar. To each 3–5 lb. of fruit use 1 lb. sugar, depending on sweetness of the gooseberries. Cover, seal, label and freeze. Cooked, sweetened and puréed gooseberries can be de-frosted and mixed with warm custard to make a fruit fool.

Grapefruit

Wash, peel and divide into segments, or slice removing the membranes and seeds. Some grapefruit have many seeds and for these it is best to cut them in halves, remove seeds then cut or spoon out sections. Pack into containers and cover with a cold 40% syrup made with the excess fruit juice and any water that is needed to make up the full amount. Leave headspace, seal, label and freeze. Store only for 2–3 months.

Grapes

Choose grapes without tough skins, wash, dry, cut in half and remove the pips. Pack into containers, without any sugar or syrup, seal, label and freeze. Alternatively, cover with a cold 40% syrup, add a piece of crumpled foil to keep the grapes under the syrup, leave headspace, seal, and freeze.

Greengages

If frozen whole, the stones tend to flavour the fruit if stored for more than a short period. It is therefore better to halve and stone the greengages and store in an unsweetened pack or partially fill a container with 50% syrup and drop in the fruit. Top up with syrup and add a piece of crumpled foil to keep the fruit under the syrup, leave headspace, seal, label and freeze.

Loganberries

See Blackberries.

Peaches

Choose firm, ripe peaches with no green colouring. Wash, remove stones and peel, preferably without immersing in boiling water. If this is necessary, do so for 30 seconds, then plunge directly into cold water. The peaches should then peel easily. Slice directly into a cold 40% syrup. To prevent discoloration, add $\frac{1}{2}$ teaspoon ascorbic acid for each pint of syrup. Press fruit down and add more syrup to cover, leave headspace. Add crumpled foil to hold fruit under the syrup. Seal, label and freeze. If you prefer, freeze in a dry sugar pack. To each $1\frac{1}{2}$ lb. fruit add 4 oz. sugar and mix well. To slow down discoloration (peaches discolour very easily) sprinkle ascorbic acid dissolved in water over the peaches before adding sugar. Use $\frac{1}{4}$ level teaspoon of ascorbic acid, dissolved in 5 teaspoons of cold water. Stir well to make sure that it is completely dissolved, before adding to every $1\frac{1}{2}$ lb. of fruit.

Oranges

For sweet oranges, see grapefruit. For Seville oranges (for marmalade) wash and freeze whole in polythene bags during the short season when they are available, for use later in the year.

Pears

Pears are not the best of fruits for freezing and you may feel that the results make it unwise to waste freezer space on them. However, for the best results, you need ripe, but not too ripe, pears with a strong flavour. Slice the pears, put into a boiling 40% syrup and cook them for 1–1$\frac{1}{2}$ minutes. Drain, cool and pack in containers and cover with cold 40% syrup. Leave headspace and add crumpled foil to keep fruit under the syrup. Seal, label and freeze.

Plums

See Greengages.

Pineapple

Choose firm, ripe pineapple with full flavour and aroma. Pare and remove core and eyes. Slice, dice or cut the pineapple into wedges or sticks. Pack tightly into containers, cover with 30% syrup made with pineapple juice or water. Leave headspace, seal, label and freeze.

Raspberries

See Strawberries or Blackberries.

Rhubarb

Choose firm, tender, well coloured stalks with good flavour and few fibres. Wash, trim, and cut into the most convenient lengths. Heat in boiling water for 1 minute and cool swiftly in cold water to help keep colour and flavour. Pack into containers without sugar, or cover with a cold 40% syrup. Leave headspace, seal, label and freeze.

Strawberries

To freeze strawberries, choose firm ripe fruit which is tart but red all the way through. The larger strawberries are better frozen sliced. Pick out the less than perfect strawberries, wash the remainder in cold water, drain and remove hulls. They may be packed dry, in dry sugar, or in a syrup pack. For the dry sugar pack, add 6 oz. sugar to $1\frac{1}{2}$ lb. strawberries and mix thoroughly. Put into containers leaving headspace, seal, label and freeze. For the syrup pack, put the fruit into containers and cover with cold 50% syrup, leaving headspace. Add crumpled foil to keep fruit under the syrup, seal, label and freeze.

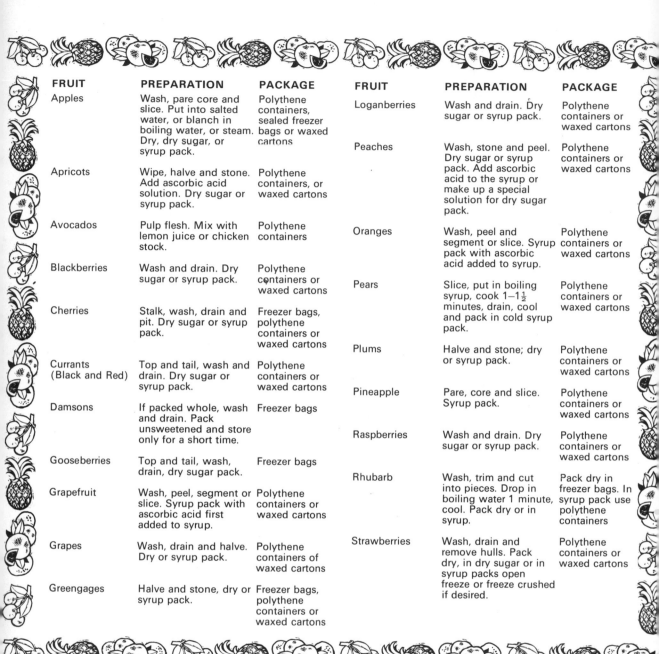

FRUIT	PREPARATION	PACKAGE	FRUIT	PREPARATION	PACKAGE
Apples	Wash, pare core and slice. Put into salted water, or blanch in boiling water, or steam. Dry, dry sugar, or syrup pack.	Polythene containers, sealed freezer bags or waxed cartons	Loganberries	Wash and drain. Dry sugar or syrup pack.	Polythene containers or waxed cartons
Apricots	Wipe, halve and stone. Add ascorbic acid solution. Dry sugar or syrup pack.	Polythene containers, or waxed cartons	Peaches	Wash, stone and peel. Dry sugar or syrup pack. Add ascorbic acid to the syrup or make up a special solution for dry sugar pack.	Polythene containers or waxed cartons
Avocados	Pulp flesh. Mix with lemon juice or chicken stock.	Polythene containers	Oranges	Wash, peel and segment or slice. Syrup pack with ascorbic acid added to syrup.	Polythene containers or waxed cartons
Blackberries	Wash and drain. Dry sugar or syrup pack.	Polythene containers or waxed cartons	Pears	Slice, put in boiling syrup, cook 1–1½ minutes, drain, cool and pack in cold syrup pack.	Polythene containers or waxed cartons
Cherries	Stalk, wash, drain and pit. Dry sugar or syrup pack.	Freezer bags, polythene containers or waxed cartons	Plums	Halve and stone; dry or syrup pack.	Polythene containers or waxed cartons
Currants (Black and Red)	Top and tail, wash and drain. Dry sugar or syrup pack.	Polythene containers or waxed cartons	Pineapple	Pare, core and slice. Syrup pack.	Polythene containers or waxed cartons
Damsons	If packed whole, wash and drain. Pack unsweetened and store only for a short time.	Freezer bags	Raspberries	Wash and drain. Dry sugar or syrup pack.	Polythene containers or waxed cartons
Gooseberries	Top and tail, wash, drain, dry sugar pack.	Freezer bags	Rhubarb	Wash, trim and cut into pieces. Drop in boiling water 1 minute, cool. Pack dry or in syrup.	Pack dry in freezer bags. In syrup pack use polythene containers
Grapefruit	Wash, peel, segment or slice. Syrup pack with ascorbic acid first added to syrup.	Polythene containers or waxed cartons	Strawberries	Wash, drain and remove hulls. Pack dry, in dry sugar or in syrup packs open freeze or freeze crushed if desired.	Polythene containers or waxed cartons
Grapes	Wash, drain and halve. Dry or syrup pack.	Polythene containers of waxed cartons			
Greengages	Halve and stone, dry or syrup pack.	Freezer bags, polythene containers or waxed cartons			

Strawberries may also be crushed before freezing. Prepare for packing in the same way, then mash. To 1½ lb. crushed strawberries, add 6 oz. of sugar and mix thoroughly. Pack into containers, leaving headspace, seal, label and freeze. Alternatively, open freeze strawberries by placing them on a tray so they are not touching. Freeze uncovered, then place in a plastic bag, seal and return to the freezer.

Fruit purées

Although you need ripe fruit for a purée, take care that it isn't over-ripe. Sort out the fruit, discarding any that is bruised. Soft fruits may be

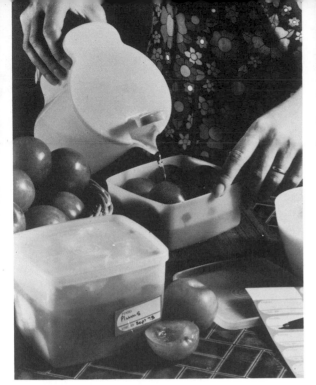

Freshly picked raspberries are open frozen on a Tupperware seal then packed in a 16-oz. square round with a foil divider between layers.

Any fruit which has a tendency to rise in sugar syrup and might discolour can be held under the syrup level with crumpled foil.

Small portions of raspberries are being packed, sprinkled with sugar, in small foil containers with transparent plastic lids which show the contents.

sieved straight away, but fruits such as plums and greengages should be put in a dish and placed in a hot oven until the juices start to run, then sieved. Sweeten the purée (using 1 part sugar to 4 parts purée, or to taste). Pack in rigid polythene containers or cartons, leaving ½-inch headspace. Seal, label and freeze. Cooked purées are treated in the same way, but must be chilled before packing.

Frozen fruit purées (or wine) can be blended with egg yolks and sugar, beaten together in a double boiler and incorporating the separately beaten egg whites to make a mousse. The mousse can be frozen in wine glasses with moulded foil caps or sealed in Tupperware dessert dishes.

Fruit juices

Prepare the fruit juice following your own favourite recipe, and allow it to become quite cold. Sweeten to taste or leave unsweetened. Ascorbic acid may also be added to prevent discoloration, though this isn't essential. Freeze the juice in rigid polythene containers or pour it into ice-cube trays or egg cups, put in the freezer until

Fully decorated parfaits can be sealed in these Tupperware parfait dishes if headspace is allowed above the piping. Here, a layer of ice cream is covered with fruit purée and piped. Seal and freeze upright before stacking.

frozen, then remove the solid cubes of juice from their containers and put them all together in a polythene bag. Seal, label and return to the freezer at once.

BREAD AND SANDWICHES

Bread can be frozen baked or unbaked and it is entirely a matter of convenience which method you choose. Storage times vary according to the type of dough, in the case of unbaked bread, and the size and type of baked bread.

Unbaked bread

The storage times for unbaked doughs are as follows:

Risen	2–3 weeks
Unrisen, plain white	8 weeks
Unrisen, enriched white	5 weeks

Doughs kept longer than the recommended times give poor results. They lose their resilience, are difficult to knock back and are slow to prove.

To freeze unrisen dough

Prepare dough until you have completed the kneading stage, then place in a large, lightly greased polythene bag. (Use the same recipe as usually.) Seal the bag tightly but leaving sufficient space above the dough to allow for any rising that may occur before the dough is frozen. Put the bag into the freezer at once.

To freeze risen dough

Prepare the dough until you have completed the kneading stage, then place in a large lightly greased polythene bag. Tie the bag loosely, leaving room for the dough to double its size. The following guide to rising time will help you.

Quick rise	45–60 minutes in a warm place.
Slower rise	2 hours at average room temperature.
Overnight rise	up to 12 hours in cold larder or refrigerator.

When the dough is risen, it will spring back from the pressure of a lightly floured finger. Turn risen dough on to a board lightly sprinkled with flour and flatten firmly with the knuckles to knock our air bubbles, then knead until firm. Place in a lightly greased polythene bag, tightly seal, and put in the freezer at once.

General points to remember

1. Freeze dough in quantities you are most likely to use. Remember that 1 lb. 2 oz. of dough makes a 1-lb. loaf.
2. Heavy duty polythene bags especially made for freezing, lightly greased and tightly sealed, are the most successful packaging for dough.
3. Correct sealing of the polythene bags is of the utmost importance as air left in the bags can cause skinning on the dough surface. The degree of skinning depends on the amount of air remaining in the bag and the storage time. Extract air by the drinking straw method (see page 49). Tighten up the seal quickly after this has been done. Even if you have allowed space in the bag for rising (in the case of unrisen doughs) you can still draw the air out of the bag in this way.

Baked bread

The length of storage depends on the crispness of the crust, but generally bread stores well for 4 weeks. Bread with any form of crisp crust stores well up to 1 week, then the crust begins to 'shell off'. Vienna and similar breads keep for up to 3 days before the crust begins to 'shell off', so do crisp rolls. Enriched bread and soft rolls store well for up to 6 weeks.

To freeze baked bread

Only freshly baked bread should be frozen. Place in an ungreased polythene bag and tightly seal. Label and freeze.

Freezing mixed packs of bought bread

A good way to ensure an interesting assortment of bread at all times is to buy several kinds of sliced bread – white, brown, rye, granary, for example. Re-pack the loaves with dividing papers between every so many slices, so that each package contains several slices of every kind. A boon when unexpected guests call and even when they don't the family will appreciate the luxury of always having a variety of different breads.

Sandwiches

There are many occasions when sandwiches are needed unexpectedly. They can be stored in the freezer for up to 4 weeks. Simply take the packed sandwiches out of the freezer and within 2–3 hours they are ready to eat.

It is always fun to try out lots of new ideas for interesting sandwich fillings.

Avoid fillings that do not freeze well – the white of hard-boiled egg, for example, mayonnaise or salad cream, raw salad foods such as lettuce, celery, tomatoes. Be cautious with jam or jelly fillings because they have a tendency to soak into the bread making it soggy and unpalatable. Indeed, it is better to butter both slices of bread right to the edge, to provide a coating that fillings find hard to penetrate.

If you need some kind of 'binding' to hold a filling together, use soured cream instead of mayonnaise or salad cream.

Open sandwiches and canapés can be made for a party up to one week beforehand. They are better laid on a foil tray and covered with moisture-vapour-proof paper. Allow 1 hour for thawing before they are served.

Making sandwiches for freezing

1. Remove the butter or margarine from the refrigerator to soften for easy spreading (but don't warm it or it may become oily).
2. Prepare all the various fillings and put in the refrigerator until you are ready for them.
3. Assemble all the packaging, sealing and labelling materials.
4. Line up the bread slices using an assortment of white, brown, rye, etc.
5. Spread each slice with butter or margarine right to the edge.
6. Put the filling on to half the bread slices, making sure you have an equal amount on each slice – use an ice cream scoop or measuring spoon if necessary. Spread the filling over the bread evenly to ensure even thawing.
7. Cover the fillings with the remaining slices of bread.
8. Stack two or three sandwiches so that you can cut through them with one stroke. Use a sharp knife so that the bread is cut, not torn.

Home-made bread dough can be frozen successfully for up to 8 weeks depending on whether the dough is risen or not. First, mix the dry ingredients together; add the yeast liquid according to your usual recipe.

Turn out on a floured board and knead for about 10 minutes.

Put the dough in a greased polythene bag. If it is to be frozen unrisen, seal loosely and draw out air with a drinking straw. Seal tightly and freeze. For risen dough, tie the bag loosely and allow the dough to rise. Then draw out air and seal tightly.

Put the dough into the freezer making sure that it is clearly labelled Risen or Unrisen dough. It is a good idea to add the date by which the dough must be used.

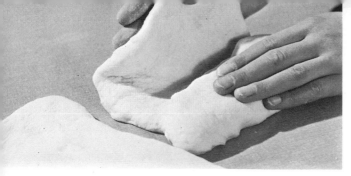

After freezing, knock back and shape the dough, remembering to allow the unrisen dough to rise first. Knock back by flattening the dough with the knuckles and folding into three.

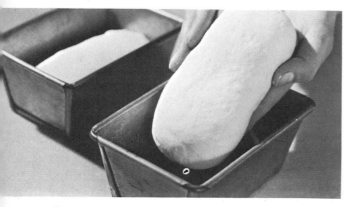

When the dough is moulded to the desired shape it can be placed in greased tins, and left to rise again. Put the tins on a baking tray and place in a hot oven.

Cut the sandwiches into sizes that are easy to handle. For children, cut them into fancy shapes with biscuit cutters.

9. Wrap immediately, either individually or in small packs using an assortment of bread types and fillings. Exclude air, seal.

10. Label clearly indicating exactly what the pack contains. If Johnny doesn't like Salami, for instance, write 'No Salami in this pack' so that he knows which one to choose.

CAKES, PASTRIES AND BISCUITS

Baked foods are most successful when frozen, so next time you are baking make extra items that you can put in the freezer. Simply follow your usual recipes for preparing and baking. Cool, then package in suitable sizes.

Cakes

Cakes can be frozen whole, or sliced. They can be iced before freezing, but pop them in the freezer unwrapped until the icing has set, then wrap, seal, label and freeze in the usual way. If you wrap before icing is set it will stick to the wrapping paper and spoil appearance of the cake when thawed.

Biscuits

Cooked biscuits are not worth the loss of freezer space because they keep perfectly well in airtight tins for many months at a time. Uncooked biscuit mixtures can be frozen. The ones containing plenty of fat and little moisture are the most successful. Pack in portions small enough to provide only enough biscuits for one serving. If preferred, slice the dough to the required thickness before freezing and separate the slices with sheets of moisture-vapour-proof paper or foil.

Pastry

Pies may be frozen baked or unbaked, but if you prefer to freeze them unbaked, don't cut vents in the top crusts until removed from the freezer just before cooking, otherwise the filling may dry out. Tarts with a custard filling are not recommended for freezing, nor are pies with meringue toppings.

Save time on busy days by having ready prepared pie tops or tart shells. Roll your pastry out and cut several circles. Place them on a cardboard ring to stop them from getting damaged. Separate each pastry circle with freezer paper or foil so that the required number of tops can be easily separated. Pop the remainder back into the freezer quickly before they begin to thaw. Pastry is usually baked unthawed, but these tops and shells will have to be thawed so that they will be pliable enough to form into the desired shape.

To wrap pies: use cling wrap or a similar material that will mould close to the pie. A stockinette 'sleeve' will help to hold the paper tightly against the pie. Seal, label and freeze.

Storage times for pies, cakes and biscuits

	Baked	Unbaked
Pies	4 months	6 months
Biscuits	6 months	6 months
	Without fat	**With fat**
Cakes	4–6 months	8–10 months

Chapter Five

Packing and storing food in the freezer

The importance of careful preparation has been stressed in Chapter Four, but no less important is the need for good air-tight packaging if the food is not going to suffer during its stay in the freezer.

Be sure to use only the materials that you know are suitable for a freezer. Many ordinary wrappings cannot withstand low temperatures and can crack, burst, or leak, allowing cross-contamination and smells to invade the freezer and ruin the food that was stored in the faulty material. Not only are wrapping materials intended to prevent moisture from escaping out of the pack, they must also protect the foods against the dry air in the freezer. Good freezer wrapping materials are therefore known as moisture-vapour-proof – a term you will come across frequently.

Use only tested wrapping materials

If you find something that looks like the perfect wrapping material, test it in the freezer *before* you use it to package food. Make up a dummy package and leave it in the freezer for a few weeks before you attempt to use it for wrapping food. It is unwise to put experimental packs containing food into the freezer, for if the material cracks or bursts you risk spoiling other food. It will probably save time and freezer space, if you buy the wrapping materials that have already been tested for the way they stand up to low temperatures. As soon as a manufacturer finds he has a suitable product for sale, he will label it accordingly.

Wrapping materials are intended to be: 1. Moisture-vapour-proof. 2. Waterproof and greaseproof. 3. Smell-free. 4. Durable and easily handled. 5. Economical on storage space. 6. Resistant to low temperatures.

The loss of moisture from frozen foods

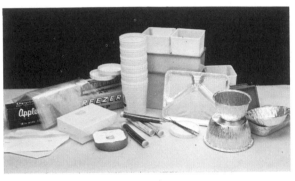

Packaging food for the freezer is a decided art, but with the help of numerous freezer-tested wrapping products that can now be bought, the job is made much easier. Shown here are some of the products available, including plastic containers, waxed tubs with screw-top lids and waxed boxes, freezer paper and polythene film, freezer tape and plastic coated bag fasteners for sealing. Also displayed are a number of aluminium foil containers such as trays, basins, pie plates and divided plates for complete dinners.

Heat-sealing a polythene bag with a Bosch electric heat sealer.

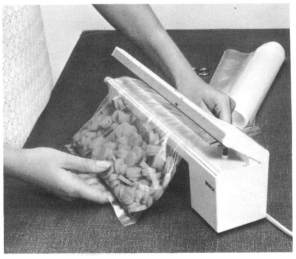

increases dehydration considerably. At low temperatures dehydration is inevitable, but if food is well wrapped with suitable materials, the loss of moisture can't take place, therefore slowing down dehydration to such a degree that it is unnoticeable for many months of storage. Contact with air is one of the worst things that can happen to frozen food during its storage period. Not only does contact with air make food dehydrate, but where air can get in a package, frost and ice particles form, causing irretrievable damage. Air makes fat become rancid more quickly.

Contamination from smells and cross-flavours can be transferred from one food to another unless packaging is done correctly.

Freezer burn

Air is an unwelcome intruder inside any pack, because it can set off a whole chain reaction starting with dehydration, and ending with freezer burn. Once the dehydration begins, oxidation of the food will take place. The effects of oxidation are seen clearly on meat and poultry in the form of mottled grey spots and these greyish marks are known as freezer burn. Only well packed foods with proper freezer-safe sealing can be kept burn-free and if a material liable to tearing or puncturing is used, precautions must be taken to protect it, by padding sharp bones or by overwrapping.

Water vapour

Since the loss of moisture is a vital factor in the spoilage of frozen food it is as well to explain more fully why this should be. At first it might appear that a home freezer, which is after all only a refrigerated box, stops the usual movements of air except when its lid or door is opened. This isn't so. Ideally, an equilibrium would be established and loss of ice from the food would cease. Indeed, it would cease if a uniform temperature could be established throughout the freezer. In most domestic home freezers, only the walls are refrigerated, though some have refrigerated shelves. The door or lid is insulated against loss of cold air, but the surface is slightly warmer than the rest of the freezer. Although

the temperature difference is slight, a small circulation of air is set up. When the air touches a colder surface it deposits frost on it, then it returns to a slightly warmer surface and takes up some of the moist vapour. This process goes on all the time.

Another way in which temperature alterations are set up, is by the differential of the thermostat – that is, when the freezer reaches a pre-determined low temperature the thermostat cuts out the action of the refrigerating unit; the temperature then starts to rise slowly to a certain degree, then the thermostat will recommence the refrigeration process. This creates a regular rise-and-fall cycle over a temperature range which is admittedly small.

The temperature also fluctuates according to general use of the freezer. The door is opened and warm air displaces cold. Unfrozen packages placed in the freezer raise the temperature and, if it is partly unloaded to reach a lower package, the packages are often returned to a different, perhaps, warmer, position.

The general effect of all these temperature fluctuations is to move ice from one part of the freezer to another and set up a decided air flow. The effect on food if it were left uncovered or badly packed and sealed, would be quite considerable. Ice would be lost from the food in the first instance, then, after another temperature change, more ice would be deposited on its surface. Unfortunately, the surface ice is not the equivalent of the ice lost from the body of the food, and dehydration, followed by oxidation and freezer burn, will result.

Packaging materials

It is almost easier to describe the kind of packaging materials to avoid, than to say exactly what you should buy. After all, the method of packing you use is largely of your own choice even though one kind of packaging will obviously suit certain foods better than another. New products come on to the market bringing with them new ideas for good and safe storage, and you can devise many 'make-do' packagings of your own, to save a little money – using empty yogurt or cream cartons, for instance. If you economise with

Chocolate and lime crunch pie (page 114) and Pineapple and mallow krispie pie (page 115)

Piquant avocado dip (page 128)

used cartons, bags or tins, do test them in the freezer before you use them to pack food.

Watch out for the following points:

1. It is no use finding and testing an ideal packing material only to find that it takes up twice as much freezer space as any other kind – that's *really* false economy! Huge round containers, for example, will leave big air spaces where it is impossible to store other packages.

2. All wrapping materials must be strong enough for the average amount of handling – not only when wrapping, but after it has been reduced to sub-zero temperatures.

Basic equipment and packaging materials

These can be improvised to a certain extent, but the following lists include useful items, some of which are everyday kitchen equipment. Labour-saving aids can be added to your collection, gradually, but they are not vital. What is vital, is the scrupulous cleanliness of all equipment and packaging.

Useful equipment for home freezing

Large saucepan 8–12-pint size	Heat sealer
	Chopping board
Fine wire mesh basket to fit saucepan *or* suitable strainer or colander	Polythene or enamel bowl to hold iced water for chilling
	Straining spoon
Selection of very sharp stainless steel knives	Tablespoon
	Fork
	Kitchen scales
	Pint measure
Sieve	Freezer thermometer

Suggested packaging materials

Foil dishes, with or without lids	Used cream cartons
Kilner jars	Moisture-vapour-proof paper
Polythene containers with air-tight seals	Polythene sheets
Polythene bags	Moisture-proof-labels
Heavy gauge sheet aluminium foil or double thickness kitchen foil	Freezer tape
	Chinagraph pencil
	Gem marker
	Drinking straws

Heat sealing sleeve polythene	Freezer cooking bags
	Cling wrap

Bag and sheet wrappings

The plastic film out of which freezer bags and sheet are made looks exactly like any ordinary polythene bag or sheet, but it is made from a heavy duty material that withstands low temperatures. Alternatively, use self-cling film which does not need sealing. Also try the moisture-vapour-proof paper such as Freezer Wrap. Bags and sheeting can often only be used once, and there is considerable danger that the food could be ruined if the plastic should tear – not unlikely

if the packages are handled often when searching for something in the freezer. Overwrapping with stockinette, greasproof paper, brown paper, or even old nylons, can help protect the wrapping from punctures.

Use sheeting for such irregularly shaped foods as meat, fish, poultry, cakes, pies. This material lends itself perfectly to the druggist's method of wrapping – a method that ensures a good air-tight seal. (See pages 38 and 68.)

Use heavy duty polythene bags for irregularly shaped foods and for the dry packaging of fruits and vegetables. Or try freezing in the special polythene bags which can be dropped straight in boiling water to heat or complete cooking.

It is a good idea to put these packages together in a large batching bag or other container – not only to help with identification (see page 71) but to protect them from damage in the freezer.

Heavy duty polythene bags can easily be used for freezing liquids. Stand the bags in rigid containers and pour in the liquid carefully. Seal the bags tightly and leave 1-inch headspace. Freeze quickly inside the containers until solid and then remove from the containers. It is then possible to store them safely in the freezer as regularly shaped solid blocks. (See illustration on page 65.)

Aluminium foil and cellophane

Ordinary cellophane is not suitable for freezing, but you can buy heat sealing cellophane which has been specially designed for freezer use. Special freezer foil is available, or use a double thickness of household foil, following the druggist's method of wrapping. Try to mould the foil closely around the food then seal by folding to eliminate all the air. Foil is most suited to wrapping meat, fish, poultry, cakes and pies. Heat sealing cellophane also lends itself to druggist's wrap, but the package should be overwrapped to protect it against puncture – stockinette is probably the best for overwrapping although brown or greaseproof paper will in many cases give adequate protection. Foil and cellophane can seldom be used a second time. Small portions of food moulded in foil can be packed together in a large Tupperware container for convenient storage.

Containers

Rigid containers are valuable in the freezer in many ways, and one of the most important features is that their regular shapes aid economy of freezer space. Inevitably, irregularly shaped packages tend to take up rather more room, because it is so difficult to pack other containers close to them without leaving unnecessarily large gaps.

Rigid polythene containers

These are the most expensive form of packaging. However, they can be used repeatedly so that after the initial outlay the cost is minimal. There are several manufacturers of polythene containers on the market, and one of these, Tupperware, claims that their products remain flexible down to 70 deg. F. below zero while the average home freezer temperature ranges only from 0 deg. F. to −10 deg. F. They withstand very hot liquids other than fats. Tupperware containers provide an air-tight seal that eliminates the need for sealing tape. Among the favourite sizes for freezing are the 20-oz. coffee canisters, and the 16 and 30-oz. 'square rounds', because they hold good family-sized portions. As the seal is so good, it is sometimes a little difficult to get the lid on the container. Use the thumbs to press down the lid on the rim of the container, beginning with the nearest side. Work round until only a slight bulge remains in the side furthest away and stretch the lid over that bulge so it snaps into place.

Polythene containers are best for foods that can be packed into them without leaving pockets of air – vegetables such as peas, for example, syrup or sugar-packed fruit, minced meat, soups, juices, purées.

In general, look for containers with these additional qualities:
1. Nest easily inside each other when not in use.
2. Stack steadily on top of each other in the freezer to avoid taking up too much space.
3. Have lids that provide a good seal.
4. Allow easy filling and labelling.

One problem with polythene containers is that food such as onions and orange juice tend to leave their aroma on the container. This can easily be removed by rinsing the emptied container in cold water; wash it in warm detergent water, and then rinse well again. Leave open to air.

Waxed cartons

These are available in various shapes and sizes, and have roughly the same uses as plastic containers. The initial cost is far cheaper, but extreme care is needed if they are to be used more than a couple of times. Freezer tape is needed for sealing the lid. Some cartons are sold with special liners and, as the carton is used only as an overwrap, just the liner needs to be sealed. Round waxed cartons, or tubs, are useful for foods such as purées, sauces, ice cream and soups; but they do take up more freezer space than the square or rectangular cartons. Some waxed tubs are provided with their own screw-top lids – perfectly adequate for the freezer without additional freezer tape.

Glass containers

Kilner jars, normally used for preserving fruit, are ideal for freezers because the seal is so good, but always make sure to leave a very generous headspace, to allow for the expansion of the food and contraction of the glass, in freezing. Narrow necked bottles are unsuitable.

Tin containers

Scout around in the kitchen for tins which would otherwise soon be discarded. Biscuit tins, for example, or syrup tins; but do take care that they are entirely made of metal and not cardboard drums with metal tops and bottoms. Square tins are the best shape in terms of freezer space, but their push-on type of lid doesn't make such a good seal. On the other hand, the press-in type of lid that comes with round tins does seal the tin well. If you can find square tins with a round press-in lid, and there are some about, you will have the ideal container on both counts.

Foil trays

Aluminium foil trays are suitable for packing plate dinners, canapés and small cakes which might crush. They are sold in a variety of sizes and shapes so that you can find exactly the right one for your requirements. Use together with a film wrapping and freezer tape. Or buy shaped foil dishes with their own lids – either opaque or see-through.

Sealing

The whole point of packaging is to provide an ait-tight container, to preserve your frozen food in the freezer – and the way in which this

Foil containers are an easy and convenient form of wrapping especially if the food such as these Cornish pasties could be easily crushed by other packing methods. Cover with polythene film by cutting a piece wide enough to go round the tray with 3–4-inch overlap, plus an extra 2–3 inches at both ends. Use the druggist's method of wrapping and fold the end pieces into a triangle.

The package is completed by sealing the two ends with freezer tape, though it could equally well have been heat sealed using an ordinary hand iron or proper heat sealer.

is usually achieved, is by the method of sealing. There are a number of ways to do this. Film wrappings can be sealed by heat, or for wax cartons a special freezer tape is needed. A twist of plastic covered wire provides a satisfactory seal for polythene bags while a screw cap or cork is quite sufficient to seal bottles. The rubber ring and metal top supplied with all Kilner jars makes a suitable seal for the freezer. Waxed tubs have their own press-in lids; the tubs can be re-used if handled with care, but it is as well to use a new lid. These are sold separately. Waxed tubs with screw-on lids need no other seal.

How to heat seal

One of the most efficient ways of sealing cellophane or plastic film wrappings is to use heat. This can be done with a proper sealer but if you haven't one an ordinary hand iron would do. Set to a warm position but don't let the film come in direct contact with the iron – put the two thicknesses of film between pieces of paper then apply the warm iron. The heat will begin to melt the wrapping material, causing it to fuse together. Remove the heat and as soon as the join is cold again, the seal will be completed.

Labelling

To get the very best out of your freezer, the packages must be carefully labelled. Not only do you need to know what is in the packages and what weights they contain, but if you are to use the packs in any semblance of order, they have to be dated as well. Plastic and wax containers need something greasy to write on them. A Colourcraft pencil will fill the bill. If the surface is warm, the greasy pencils will write better.

Ordinary labelling materials just can't cope with the sub-zero temperatures of a freezer but special products that can have been evolved. Instead of pens with water-based ink, for example, a chinagraph pencil or gem marker is needed. Also shown here are neat self-adhesive labels which will adhere tightly to most surfaces. Note the handy polythene clips used to seal the bag of peas.

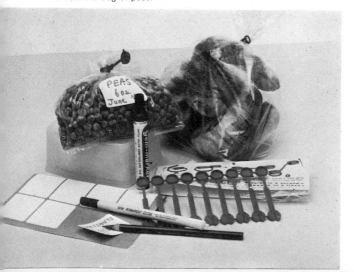

Small adhesive labels can be written on with a chinagraph pencil or gem marker (ink smudges in the freezer) and can be easily stuck on to all kinds of packaging materials.

Methods of packing

Even the best of packing materials will not do the job adequately if the actual method used is not a good one. Materials can be completely moisture-vapour-proof, but if they are not wrapped and sealed properly they will let air in and moisture out. There are two recommended ways of packing the sheet wrapping – one is Druggist's Wrap, the other Butcher's Wrap.

Druggist's wrap

This is the most used of the two wrappings. Put the food in the centre of your material, allowing enough to cover the food plus a 3–4-inch overlap. Bring the longest edges together over the food and, working away from you, fold the two edges over about 1 inch. Now, fold over and over until edges are tight against package. Press out all air pockets and fold the end flaps over to make triangles and tuck ends under the package. Seal by heat or with freezer tape. Overwrap fragile packages.

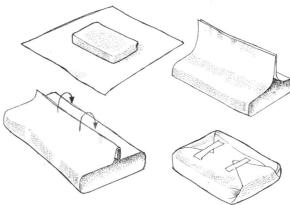

Butcher's wrap

Place the food in the corner of your chosen wrapping paper and fold that corner over the food. Fold the two side pieces across the top, then roll the package over and over to the end of the paper. Fasten by heat or tape. Overwrap fragile packages. (See illustration page 69.)

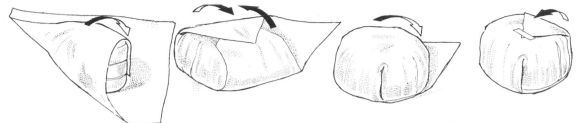

Headspace

The water contained in foods will expand on freezing so this must be taken into account when packing. The aim is to achieve a full container after freezing so just enough space is left at the top to allow for the expansion – too little space may cause the container to burst when the contents expand, spilling them into the freezer; too much space will allow dehydration to take place. It is obvious that a certain amount of judgement must be exercised if you are going to leave the right amount of headspace, but as your experience grows, so will your judgement! It is better to leave too much headspace than too little. The following will give you an approximate guide:

Pack	Headspace per pint
Dry pack	$\frac{1}{2}$–1 inch
Wet pack (narrow top)	$\frac{3}{4}$–1 inch
Wet pack (wide top)	$\frac{1}{2}$–1 inch

For packs larger than the usual pint size, leave double the recommended headspace.

Strong sugar solutions will freeze at progressively lower temperatures, not at a precise, high temperature as in the case of water. This means that while water, stored in glass bottles, may burst the containers, the syrupy liquids just expand into the necks. So, with the appropriate headspace allowed for, bottles make a suitable container for sauces and fruit juices.

HANDY PACKING HINTS

1. Wrapping pastry cases

Freeze large uncooked pastry cases by placing the first on a cardboard base, covering it with a foil divider and putting the next case on top and so on, until you have a stack ready. Open freeze and, when hard, put into a polythene bag. It is then quite easy to peel off a case each time one is required. Small tartlet cases can be open frozen in patty tins, then packed inside one another before placing in a polythene bag. Seal in the usual way.

2. Packing fruit in syrup

Many fruits are recommended to be frozen in a syrup solution rather than by the dry pack method. The surface fruit is liable to swim to the surface of the solution. To keep it submerged, leave the usual headspace, but crumple a piece of foil and place it between the fruit and the lid.

3. Iced cakes

Open freeze iced cakes on a baking tray. Lay a long pad of foil across the bottom of a rigid based container before putting the cake on top. Leave enough foil at each side to lift it out when thawed.

Saving freezer space

When space in the freezer is short, a lot can be done to help by using packaging designed to save space. If you are on a really stringent space economy drive, use square or oblong containers. Close-packing enables you to get far more into the freezer. But remember, close packing when freezing down, delays freezing. And don't pack so closely that you have no finger spaces to allow you to get a package out. For if the packages have frozen together you could even need a chisel to separate them!

A round container stops close-packing but, if the top is flat, more containers can be stacked on it. Tubs that are wider at the top than at the bottom waste space; so too, do odd shapes. Generally speaking, a proportion of one third rigid based containers to two thirds bags or other packs, gives more economical storage than all bags; and the more containers you can afford, the better you can exploit your freezer space. A combination of round and square round shapes gives adequate finger spaces. Foods like chicken patties can be easily moulded in foil and closely packed in a large polythene container.

The space that is saved, or wasted, with plastic bags, depends entirely on the shape of the food packed in them. Such items as poultry or shoulder of lamb make awkward packages and waste more freezer space than a bag of peas.

Making up dummy packs to test the freezer-worthiness of packaging materials must be postponed when space is limited, and it is not worth devoting valuable room to foods that do not freeze well. Good packaging also plays a considerable part in saving space. Not only does a neatly and efficiently wrapped parcel prevent the unwanted air from creeping in and damaging the food, it also means more space for food.

Keeping a record

Not only is it important that the packages of frozen food are labelled correctly, it is wise to keep an up-to-date record of everything you put in the freezer. If you remove one of three frozen beef casseroles, for example, you must then record the fact, for you can hardly remember from one month to the next exactly what the cabinet holds. The easiest method, and the best to see at a glance, is a chart record with columns for the type of food, amount in each package, date frozen, where it can be found in the freezer and any other useful information you may wish to include. Hang the chart on the freezer itself or on a hook nearby for the greatest convenience.

Freezer Log			
Date in	Description of pack	No of packs	Shelf or basket
10 June	Apricot purée – Square Rounds	///// XXX	4 – right hand
2 July	Blanched peas – Green poly-bags	// XXXX	2 – centre
4 July	Gooseberry Fool – Tupperware Tumblers	/////	Bottom – left hand

Using colour for identification

Visual aids to recognising the contents of containers without being able to read the labels are always helpful. For instance, when you add to your stock of bought containers, vary the shapes, sizes and colours, if possible. Rigid polythene containers of different colours help. Pack a batch of peas in square green Tupperware containers, and a batch of runner beans in square blue ones, for example. Reserve similar pink and yellow containers for strawberries and raspberries, or other soft fruits.

If your containers are all white, make it a rule to use round ones for fruits, square ones for vegetables. This simplifies recognition of packs you can't easily touch, in a hard-to-reach corner of the freezer. Plastic coated wire closures come in many different colours and, provided you note on your chart which colour you use on any particular batch, they help to pick out the bag you want. A helpful innovation is the introduction of coloured polythene bags. You could use green, say, for vegetables and blue for fruit, making selection from the freezer simplicity itself. These are opaque, but if you particularly like to see the contents in each package, get the new stripy bags. The alternate stripes in transparent and coloured polythene allow you to see exactly what is in the pack.

Always record these identification aids very clearly on your chart. You may think at the time that you will easily remember which set of packs went into the freezer first, but this becomes increasingly difficult, if not impossible, as the months go by.

Another useful investment is a number of coloured polythene batching bags, large enough to take several smaller packs and a marvellous aid to quick identification. Devise the colour code that best suits your needs, but you could, for example, keep all your soft fruit in a red batching bag and your green vegetables in a green bag, and so on. The packs inside the bag can be arranged to fit the storage space available, even if it's an awkward one, as the bag accommodates itself to any shape you need. This is the big advantage these bags have over the cardboard-box method of batching.

The chart serves another very useful purpose. It is a record of your freezing needs. When it is full, put up a new one, and study the old one. It shows how many pounds of fruit and vegetables you have frozen and used during the period it covers; how much cooked food you prepared and stored, how quickly it was used up, and whether any items 'hung fire' and were not worth giving freezer space. You can also calculate how much money you save on buying meat or other items in bulk at special low rates. These old records are valuable, especially when planning your freezing needs for the coming year.

Chapter Six

Thawing out frozen food for use

The rules about thawing are simple, but they must be borne in mind, for it is a pity to spoil successfully frozen food for want of proper de-frosting.

It isn't always necessary to completely thaw food before cooking; indeed, most vegetables may be cooked straight from the freezer. You sometimes need to part-thaw food so that you can remove the dividing wrappers between the layers, or to remove the food from its container. To do this place the package in lukewarm water for a few minutes or leave it at room temperature for ½–1 hour. Prepared food that is to be cooked over direct heat (such as fish cakes) may need a little thawing first.

Food containing a high proportion of fluid, such as soup, should be heated in a saucepan over a low, direct heat. Stir occasionally, to prevent the food sticking. Other dishes may be heated in a hot oven or in a double boiler. Heating in the oven needs almost no attention but it does take longer. For double boiler heating, place warm water in the lower pan and, as the frozen food begins to thaw, gently break it up. Stir when necessary to prevent sticking, but don't stir too often or you will alter the texture of the food.

To speed up the thawing of large blocks of frozen food, place in a wide bottomed pan set in a meat tin of hot water. Remember that food which takes longer than 3–4 hours to thaw at room temperature is potentially dangerous. It is better thawed in the refrigerator. Once de-frosted, reheat or cook quickly and use at once.

FRUIT

The final quality of frozen fruits can be controlled in thawing but this is one of the least recognised facts. It should always be de-frosted in its sealed container which should be turned over several times to ensure even thawing. The time this takes varies slightly, according to the way it is packed and the method you use to thaw. Fruits packed in syrup, for example, take longer than those packed in dry sugar, while unsweetened, frozen fruits will take longer to thaw than syrup packed fruits.

Fruit may be thawed in three ways; the refrigerator which gives the most uniform thawing; at room temperature; or, when time is really short, for 20–45 minutes in cold running water. The method is, therefore, determined by the time you can afford to wait.

Fruits to be eaten raw are at their best served while still slightly iced, that is frappé. If the fruit becomes too warm before you are ready to eat it, pop it back in the freezer for a minute or two.

Do not open the fruit containers until you are ready to serve or cook it, for it begins to lose colour and flavour as soon as it meets the air. Thaw only as much as you need at one time; but if any fruit is left over, heat to near boiling point, to slow down the enzyme activity, cool, store in the refrigerator and use as soon as possible.

For fruit that is to be cooked, remember to take it from the freezer early enough to allow part-thawing (until the pieces of fruit can be separated), then cook as if it were fresh fruit. Be cautious about adding sugar to the fruit when cooking it since sugar-packed fruits will already be quite sweet.

Frozen fruit often yields more juice than fresh, so if you're making a fruit pie or some other baked product, use only part of the juice or add a little cornflour to thicken the extra juice. If by some chance there is not enough juice to prevent it burning in the pan, add a little water. Experience will tell you how much.

While most fruits like slow thawing, those subject to discoloration may be helped by treating in a different way. If it is unsweetened fruit, separate and dip immediately into a hot syrup. On the other hand, sweetened fruit should be upturned in its container to make sure that the sugar or syrup completely covers it.

Strawberries need special care. De-frost in the refrigerator and serve while still chilled as complete de-frosting nearly always spoils the texture.

The following chart provides a guide for thawing fruit at room temperature and in the refrigerator. It must be remembered, however, that the depth of the container and the temperature are important factors. Fruit which is to be eaten uncooked, should be thawed but still chilled. Fruit which is to be cooked for pastries and pies, etc. need not be completely thawed. For the purpose of the chart it has been assumed that 1-lb. packs of fruit are to be thawed completely and room temperature is assumed to be around 70 deg. F.

THAWING TIME-TABLE

Fruit	Approx. refrigerator thawing time	Approx. room temperature thawing time
Soft fruits (skinless)		
Blackberries	6–7 hours	2–3 hours
Raspberries	6–7 hours	2–3 hours
Strawberries	6–7 hours	2–3 hours
Soft fruits (with tough skins)		
Apricots	7–8 hours	3–4 hours
Cherries	7–8 hours	3–4 hours
Cranberries	7–8 hours	3–4 hours
Currants	7–8 hours	3–4 hours
Damsons	7–8 hours	3–4 hours
Figs	7–8 hours	3–4 hours
Gooseberries	7–8 hours	3–4 hours
Grapes	7–8 hours	3–4 hours
Grapefruit segments	7–8 hours	3–4 hours
Peaches	7–8 hours	3–4 hours
Crisp fruits		
Apple slices	7–8 hours	3½–4 hours
Pear slices	7–8 hours	3½–4 hours
Rhubarb	7–8 hours	3½–4 hours
Fruit purées		
Apple purée	6–8 hours	2–4 hours
Avocado purée	6–8 hours	2–4 hours
Banana purée	6–8 hours	2–4 hours

VEGETABLES

Properly frozen vegetables are full of flavour, tender when cooked and ensure good out-of-season varieties all year round. The usual thawing procedures just aren't applied to frozen vegetables, for most of them are cooked while still frozen so that their garden-fresh flavour and good colour are preserved. Exceptions include broccoli and asparagus which will heat more evenly if thawed just enough to separate before being plunged in boiling water. Corn on the cob should be completely thawed before cooking so that the cob is heated through by the time the corn is cooked. It must be eaten as soon as possible after thawing and cooking since delay causes it to become soggy. Beetroot should be thawed, left covered, in a warm room for some 2–3 hours. Mushrooms, too, are left in their container with the lid loosened, for 2–3 hours.

THAWING TIME-TABLE

Vegetables	Approx. refrigerator thawing time	Approx. room temperature thawing time
Beetroot, young	9–10 hours per lb.	2–3 hours
Mushroom, caps	6–8 hours per lb.	2–3 hours
Peppers (sliced)	24 hours per lb.	1½ hours
Potatoes	6 hours per lb.	1½ hours

When cooking still-frozen vegetables, bear in mind that most of them have been part-cooked before freezing during the blanching process and that freezing softens the tissues. This means that the cooking time should be reduced by $\frac{1}{2}-\frac{2}{3}$ of the normal time of the fresh variety.

Cook only the amount of vegetables that you will eat at one meal. Once thawed and cooked, the vegetable loses Vitamin C and much of its quality. If you remember that a pint of frozen vegetable will yield four average servings you can quickly estimate how much you need for your own particular requirements.

How to cook frozen vegetables

The most common way to prepare frozen vegetables for eating is by cooking gently in a little boiling water. But this is by no means the only way they may be prepared. Other ways include steaming, baking, sautéing and the pressure cooker method. Part-thawed, cooked new potatoes should be reheated in butter slowly until completely thawed and hot right through. Alternatively they can be sautéed.

Most vegetables can be cooked while still frozen, but it is as well to part-thaw leafy vegetables such as spinach, just enough to separate the leaves before cooking. Alternatively, cut the still frozen block into 1-inch cubes before heating; this method, or part-thawing, ensures that the greens will cook more uniformly and that they have a better texture. Vegetables such as broccoli and asparagus are part-thawed before cooking.

Boiling frozen vegetables

The amount of water used for cooking the vegetables depends on the size of the package. The best guide is to allow about $\frac{1}{4}$ pint of water and $\frac{1}{2}$ teaspoon of salt to each pint package – the frost in the package will provide extra moisture. (You can even cook without extra water if you melt a little fat in the pan, cover, and allow vegetables to cook gently in their own moisture.) Put the water in a shallow saucepan with a well-fitting lid, add the salt and bring to the boil. Put the frozen vegetables in the boiling water, cover, and bring water quickly back to the boil. It is sometimes necessary to separate the vegetables carefully with a fork to ensure even cooking. Once the water is back on the boil, lower heat and measure the cooking time from that moment.

It is possible to give you approximate cooking times but the general principle is to cook the vegetables until tender, but still firm. The exact time varies with the sizes of the pieces, the variety, and the age of the vegetables. When ready, drain, and adjust the seasoning if this is necessary, add a knob of butter or a tablespoon of cream and serve immediately. Any delay before they are eaten will cause loss of quality.

The following time-table will give you an approximate guide for cooking various frozen vegetables using the boiling water method.

Vegetable	Cooking time after water returns to boil
Asparagus	5–10 minutes
Artichokes (globe)	5 minutes
Aubergine	5 minutes
Beans (broad)	8–10 minutes
Beans (French, whole)	7–8 minutes
Beans (French, sliced)	5–6 minutes
Beans (runner, sliced)	7–8 minutes
Broccoli	5–8 minutes
Brussels sprouts	5–8 minutes
Cabbage	5–8 minutes
Carrots	8–10 minutes
Cauliflower	8–10 minutes
Corn on the cob	8–10 minutes (thaw before cooking)
Peas	5–8 minutes
Spinach	5–8 minutes in butter only
Turnips	8–12 minutes

Steaming frozen vegetables

Part-thaw the vegetables so that the pieces can be separated, then put them in a steamer over fast boiling water. Cover and cook until just tender. The cooking time commences as soon as the vegetables are put in the steamer. You may use the time-table for cooking vegetables by the boiling water method as an approximate guide for steaming vegetables. When cooked, drain, season and serve at once.

Baking frozen vegetables

Part-thaw vegetables so that the pieces can be separated and place in a greased casserole. Season, cover and bake in a moderate oven (350 deg. F. – Gas Mark 4) for about 45 minutes until tender. If the oven is being used for other food at the same time, the cooking time may be slightly longer. To bake corn on the cob, part-thaw the ears, brush with melted butter or margarine, season with salt and bake at 400 deg. F. – Gas Mark 6, for about 20 minutes.

Sautéing frozen vegetables

Using a frying pan with lid, put a tablespoon of fat in the pan. Add 1 pint of frozen vegetables which have been part-thawed to separate the pieces. Cover and cook until tender over a moderate heat, stirring occasionally. Season and serve at once.

Pressure cooking frozen vegetables

Part-thaw vegetables, put $\frac{1}{2}$ pint of water and $\frac{1}{2}$ teaspoon of salt into pressure cooker and bring to the boil. Add vegetables, fasten lid and bring to 15 lb. pressure. Time vegetables while under pressure and bring pressure down as quickly as possible afterwards. Season and serve at once. Here is a time-table for pressure cooking frozen vegetables.

Vegetable	Pressure cooking time (once pressure is raised to 15 lb.)
Asparagus	1 minute
Beans (broad)	1–1½ minutes
Beans (French, whole)	1 minute
Beans (French, sliced)	½ minute
Beans (runner, sliced)	½ minute
Broccoli	45 seconds
Brussels sprouts	1 minute
Carrots	1–1½ minutes
Cauliflower	½ minute
Corn on the cob	2–3 minutes (thawed)
Peas	½ minute

FISH

Frozen fish may be thawed, still wrapped, in the refrigerator or at room temperature; and the thawing time will depend on the size and thickness of the fish and how solidly it is packed. It takes somewhere between 6–10 hours to thaw out a 1-lb. pack of fish in the refrigerator, and 3–5 hours to thaw out the same piece of fish at room temperature. Fish that is to be coated in breadcrumbs and fried should be thawed completely first, for breadcrumbs will not adhere to frozen fish. Fish that should be thawed, unwrapped, are bloaters, kippers, smoked haddock and salmon. Thaw in the refrigerator or in a larder protected from flies – and pets! Most other fish, however, need not be thawed at all. Simply increase the cooking time by about half as much again for fresh fish. Once fish is thawed, use at once. Do not re-freeze thawed fish if it smells at all strong.

Cooking frozen fish

To bake: put whole fish thawed in the refrigerator into a buttered baking dish. Fillets and steaks may be put in the oven while still frozen.

To fry: part-thaw small fish and shellfish to separate, then fry in shallow or deep fat.

To boil: rinse in cold water and put into the seasoned liquid in which you are cooking it.

To grill: part-thaw to separate the pieces and place on a greased or oiled grill pan.

To poach: bring seasoned milk or water to simmering point and drop in the frozen fish.

Shellfish

Oysters are thawed in their unopened container. This takes about 8 hours in the refrigerator and 4–6 hours at room temperature. Oysters to be eaten raw are served while some ice particles still remain. Scallops must be thoroughly thawed before they are cooked. Shrimps may be thawed or placed in boiling water to cover, add a tablespoon of salt, bring water back to the boil then lower heat and simmer for 5 minutes. Plunge the shrimps into a pan of cold water so that they may be shelled easily. Treat as fresh shrimps.

THAWING TIME-TABLE

Fish	Approx. refrigerator thawing time	Approx. room temperature thawing time
Haddock	6–10 hours per lb.	3–5 hours per lb.
Halibut	6–10 hours per lb.	3–5 hours per lb.
Herring	6–10 hours per lb.	3–5 hours per lb.
Mackerel	6–10 hours per lb.	3–5 hours per lb.
Salmon	6–10 hours per lb.	3–5 hours per lb.
Turbot	6–10 hours per lb.	3–5 hours per lb.
Trout	6–10 hours per lb.	3–5 hours per lb.
Cod	6–10 hours per lb.	3–5 hours per lb.
Plaice	6–10 hours per lb.	3–5 hours per lb.
Sole	6–10 hours per lb.	3–5 hours per lb.
Whiting	6–10 hours per lb.	3–5 hours per lb.
Crab	10–12 hours per pint carton	3 hours per pint carton
Lobster	10–12 hours per pint carton	3 hours per pint carton
Oysters	8 hours	4–6 hours
Scallops	8 hours	4–6 hours
Shrimps	6–8 hours per 4 oz. pack	1½–2 hours per pint carton
Kippers	5 hours	2–2½ hours
Smoked haddock	5 hours	2–2½ hours
Bloaters	5 hours	2–2½ hours

MEAT

It is possible to cook some meat before it is thawed. Thin tender cuts of meat are the most suitable for cooking when frozen. The latest theory is that meat retains more of its natural juices if cooked from the frozen state, but there is a danger that, while meat looks cooked on the outside, it is still raw in the centre. The same applies to meat that is placed in hot water to speed thawing; the centre of the joint often remains frozen and uneven cooking results. If you prefer to roast joints from the frozen state it is a rough guide to allow ¾ hour per lb. for beef and lamb and 1 hour for pork. And it is sensible to use a meat thermometer to check that the centre of the roast is cooked.

Thawing in the refrigerator is probably the ideal method, because a lot of the juices are re-absorbed by the meat, over the long thawing period. Although ideal, it is not always practical to use the refrigerator for thawing purposes. If a joint is required on the day of removal from the freezer, or if refrigerator space is limited, thaw in a cool place, preferably where the temperature is not over 60 deg. F.

The time taken for a piece of meat to reach the temperature at which ice crystals in the muscles start to melt (28 deg. F.), is two to three times longer in the refrigerator than at room temperature. The time taken for meat to reach 39–40 deg. F., or thaw completely, is approximately four to six times longer in the refrigerator than at room temperature.

Use frozen meat as soon as it is thawed or, if it must be stored, put loosely covered in the coldest part of the refrigerator and use within a few days. Minced meat and offal are especially perishable and should be used quickly.

To thaw lamb or beef, leave it in the refrigerator overnight still in its polythene bag or whatever packaging you have used. Alternatively, leave it covered in the kitchen for the length of time suggested in the thawing chart.

To thaw pork or veal, remove from its covering and leave in the refrigerator or, in the case of veal, allow it to thaw in a cool larder uncovered but well-protected from flies (a special wire mesh cover sold for the purpose is ideal). Pork should be thawed only in the refrigerator because of the added risk of contamination with this type of meat when thawed at higher temperatures.

Thaw *liver* uncovered. *Meat that has been cooked* before freezing and which contains no gravy or sauce, should be thawed in its sealed container. *Stews and casseroles* may simply be unwrapped and placed in a suitable dish in a low oven, or in a double boiler over a low heat. *Game* may be thawed unwrapped in a refrigerator or cool larder. Don't leave it too long after thawing because it will quickly go bad.

The time-table below gives you a guide to help you judge the thawing times for meat.

THAWING TIME-TABLE

Meat	Approx. refrigerator thawing time	Approx. room temperature thawing time
Beef	5 hours per lb.	2 hours per lb.
Veal	5 hours per lb.	2 hours per lb.
Lamb	5 hours per lb.	2 hours per lb.
Pork	5 hours per lb.	Thaw in refrigerator only as de-frosting in a warm room can be dangerous.
Sausages	6 hours per lb.	1½–2 hours per lb.
Chops (½ inch thick)	6 hours per lb.	1–2 hours per lb.
Steaks (½ inch thick)	6 hours per lb.	1–2 hours per lb.
Steaks (1 inch thick)	8 hours per lb.	2–3 hours per lb.
Steaks (over 1 inch thick)	8–10 hours per lb.	3–4 hours per lb.
Poultry (whole)	See chart opposite for thawing poultry	
Chicken (portions)	5–6 hours per lb.	1 hour per lb.
Minced meat	6 hours per lb.	1–1½ hours per lb.
Meat patties	6 hours per lb.	1 hour per lb.
Offal		
Sheep's heart	8–9 hours per lb.	1–1½ hours per lb.
Lamb's kidney	8–9 hours per lb.	1–1½ hours per lb.
Sliced liver	8–9 hours per lb.	½–¾ hour per lb.
Sliced sweetbreads	10 hours per lb.	¾–1 hour per lb.
Tripe	10–12 hours per lb.	1–1½ hours per lb.
Lamb's tongue	10–12 hours per lb.	1–1¼ hours per lb.
Game		
Hare	5–6 hours per lb.	1½–2 hours per lb.
Rabbit	5–6 hours per lb.	1½–2 hours per lb.
Cooked meat, stews and casseroles	8–10 hours per lb.	1–1½ hours per lb.

POULTRY

The general principles of thawing poultry are much the same as thawing meat, but there are a few points worth mentioning. Thawing a whole large bird can take quite a long time in the refrigerator though it is the best way if you can spare the time. Remove packaging, and leave on a tray in the refrigerator until the bird is pliable. To speed up a little, first immerse the bird in its tight bag in cold water. Change water often. Then return to refrigerator. Alternatively, leave it in its wrapping and part-thaw in the refrigerator first, then finish in cold water.

Thawing time-table for whole birds

Type	Thawing in the refrigerator
Chickens	
4 lb. and over	1–1½ days
Under 4 lb.	12–16 hours
Ducks	
3–5 lb.	1–1½ days
Geese	
4–14 lb.	1–2 days
Turkeys	
18 lb. and over	2–3 days
Under 18 lb.	1–2 days

To thaw poultry portions. Unwrap the package and separate the portions so air can reach each piece. Leave in the refrigerator to thaw – this will take between 3–9 hours depending on the size of the portions. If the portions are wanted for stewing they may be cooked without thawing, allowing a little extra time.

To thaw cooked poultry. Poultry frozen without gravy or sauce may be thawed in the refrigerator, still wrapped. Part-thaw poultry frozen in gravy, until it can be removed easily from its package, place in a pan and heat in oven (400 deg. F. – Gas Mark 6), or on top of the cooker starting with a low heat. (To speed the part-thawing, for removal from containers, dip the package in hot water.)

CAKES, PASTRIES AND BISCUITS
Cakes

Cakes should remain wrapped while thawing is taking place, with the exception of iced cakes. These are unwrapped first so that, when they have thawed, the softened icing does not stick to the wrapping and spoil the appearance of the cake. Thawing times depend on the type of cake, its size and thickness and whether it is iced or plain. Here is a general guide:

Type of cake

8-inch layer cake

1 layer, not iced	1–1¼ hours
2 layer, with ½-inch icing	2 hours
2 layer, with ¼-inch icing	1–1¼ hours

Angel cake, sponge

whole cake	3–4 hours
wedges	1¼–1½ hours

Cupcakes

	12–25 minutes

Pastry

Cooked pies that are to be served hot are placed unthawed into centre of a hot oven (400 deg. F. – Gas Mark 6) for 30–40 minutes. Cooked turnovers and small portions of pie are first unwrapped then, using the same temperature, put in oven for 10–15 minutes.

Uncooked pies are not thawed before baking. First remove wrappings, then split vents in the pastry top and bake in a hot oven (400 deg. F. – Gas Mark 6), until the pie is brown. This will probably take about 15 minutes longer than a pie that has not been frozen. If the filling in the centre of the pie is bubbling you'll know that it is heated right through.

Cooked pies that are to be eaten cold may be thawed in the refrigerator, or at room temperature. A 9-inch pie will thaw at room temperature in 1½–2 hours.

Biscuits

If you do freeze baked biscuits, thaw at room temperature in their wrapping or container. If they are thin biscuits they will thaw in 15–30 minutes. When only a few biscuits are required, remove them from their container and wrap them in some other suitable wrapping before allowing them to thaw; then return the rest of the biscuits to the freezer. For unbaked biscuit dough, remove from the freezer and part-thaw until dough can be sliced, then bake in the usual way allowing a little longer than for biscuits that have not been frozen.

DAIRY FOODS

Eggs

Having already packed eggs for savoury or sweet mixtures, take care that you are thawing the right package. Thaw in the refrigerator, at room temperature, or under cold running water. They may then be used as fresh eggs, though allowance must be made for the salt or sugar already in the egg when adding other ingredients. Here is an approximate chart of equivalents to help you:

Approx. 10 whole large eggs = 1 pint
Approx. 24 large egg yolks = 1 pint
2 tablespoons thawed yolk = 1 egg yolk
3 tablespoons thawed whole egg = 1 egg

Other dairy products

Most dairy products including butter, milk, cheese and cream, are best thawed in the refrigerator for absolute safety. The exceptions are little mounds or rosettes of whipped cream ready to pop on top of individual desserts. These may be served still frozen since they thaw very quickly.

THAWING TIME-TABLE

Dairy foods	Approx. refrigerator thawing time	Approx. room temperature thawing time
Eggs, whole	18–20 hours per ½ pt.	1½ hours
Eggs, yolks	18–20 hours per ½ pt.	1½ hours
Cream	8 hours per ¼ pt.	1½ hours
Cheese	8 hours per ¼ lb.	2–3 hours
Butter	2 hours per ½ lb.	30 minutes
Milk	18–20 hours per pt.	1½ hours

BREAD AND SANDWICHES

Uncooked bread

It takes 5–6 hours to thaw uncooked bread doughs at room temperature or they may be de-frosted overnight in the refrigerator. Don't leave overnight at room temperature or the dough will be over-risen. It should be thawed in its polythene bag to prevent a skin forming, but first unseal the bag, then re-seal loosely at the top to allow space for expansion when the dough rises.

Baked bread and rolls

Leave to thaw in a sealed polythene bag at room temperature, or overnight in the refrigerator.

To crisp the crust, remove wrapping when thawed and place bread in a hot oven (400 deg. F. – Gas Mark 6) for 5–10 minutes. Individual slices may be thawed conveniently in a toaster or under the grill using a low heat for a short time.

Sandwiches and canapés

The exact thawing time of sandwiches will depend on the thickness of the bread slices and the type and thickness of the filling, but a pack of wrapped sandwiches will take 2½–3 hours to thaw at room temperature.

Canapés can be thawed, wrapped, in the refrigerator in 1–2 hours, or unwrapped and placed on serving plates to thaw out in 15–20 minutes. Serve while still cold. If large quantities of canapés are wanted, don't remove from the freezer all at once, but thaw them as needed.

THAWING TIME-TABLE

Bread and sandwiches	Approx. refrigerator thawing time	Approx. room temperature thawing time
Uncooked bread dough	Overnight	5–6 hours
Baked bread	Overnight	2–3 hours
Baked rolls	Overnight	1 hour
Baked bread slices		2–3 minutes in toaster or under grill
Sandwiches		2½–3 hours
Canapés	1–2 hours	15–20 minutes

Soups and sauces

Clear soups may simply be thawed in a pan over a low heat. On the other hand, cream soups should be thawed in a double boiler and, if they separate, beaten vigorously. Sauces of all kinds, savoury and sweet, are thawed in the same way as cream soups.

Fruit purées and juices

Like the fruits themselves, fruit juices and purées may be thawed in any one of three ways – the refrigerator which gives the most uniform thawing, at room temperature, or, when speed is essential, in cold running water for a few minutes. Don't open the container until you are ready to serve or heat through, and thaw only the amount you require at any one time. For this reason it is recommended that fruit juices and purées are frozen in small convenient portions.

THAWING TIME-TABLE

Fruit purées ½ pint packs	Approximate refrigerator thawing time	Approx. room temperature thawing time
Apple purée	6–8 hours	2–4 hours
Avocado purée	6–8 hours	2–4 hours
Banana purée	6–8 hours	2–4 hours

To thaw fruit juice allow 2–3 hours at room temperature or 4–6 hours in the refrigerator.

When can you re-freeze?

This chart applies only to food that has partially thawed. Soft ice crystals must remain, throughout the food, for absolute safety. If it has been thawed for a prolonged or unknown time, don't re-freeze on any account.

Frozen uncooked food	Re-freezing
Fruits	Yes
Fruit juices and purées	Yes but flavour may deteriorate and reconstituted juice may separate.
Vegetables	Yes
Meat	Yes
Offal	Yes, unless it has an abnormal odour, when it should be discarded.
Poultry	Yes, unless it has an abnormal odour, when it should be discarded.
Fish	Yes, unless it has an abnormal odour, when it should be discarded.

Frozen pre-cooked foods	
Meat, poultry and fish pies	No
Complete made-up dishes such as casseroles	No
Soups	No
Fruit pies	Yes
Ice cream	Yes, but texture will be affected. It may help if you can beat once or twice during re-freezing.

Should you re-freeze?

There are varying opinions on how far one may safely go in re-freezing thawed food. In every case it is better to eat thawed food as soon as possible and if there is any risk attached to re-freezing the remainder, DON'T. This certainly applies to completely thawed food that has remained at room temperature for even a short space of time.

Re-freezing causes loss of quality, flavour and, probably, nutritive values too. When the food was originally frozen, bacterial growth was arrested, not killed. This means that the growth begins again as soon as the food is thawed, and has reached a more advanced stage by the time you want to re-freeze it.

There are two safety factors to bear in mind when deciding whether or not to re-freeze.
1. Was only the best quality food chosen for freezing and was the original preparation completed under the most scrupulously clean conditions? If the food was not well selected and handled, the bacterial count will be much higher and spoilage will develop very quickly after thawing. This applies to the more highly perishable foods such as meat, fish, poultry and meat pies. If there is the faintest suspicion that the food is unsafe to re-freeze, it really is not wise to take the risk.
2. The other factor is the length of time the food has been thawed or part-thawed before re-freezing. The chart opposite will help you to decide what foods to re-freeze.

All the well-known frozen food companies advise their customers not to re-freeze their products, and this is the best general rule, whether the food has been frozen commercially or by yourself. But it *is* possible to re-freeze safely and only you can decide whether you want to take the risk.

Another point to remember – any cooked or made-up dish that has been prepared from frozen raw meat can safely be frozen, since this is regarded as a first freezing for the completed dish. The re-freezing chart on the left is based on the assumption that the food was in prime condition and handled under hygienic conditions.

Part Two
Wonderful meals from your freezer

Chapter Seven

Cooking ahead — the 'eat one, freeze two' plan

This section is devoted entirely to providing food for the day-to-day needs of the family. All of the king-size recipes found here are intended to give three complete meals, each one large enough to serve a family of four people. They can easily be adapted to meet your own particular needs – increasing the amounts by half as much again, for example, if yours is a family of six.

Cooking in bulk

The most workable plan seems to be to cook your main meal requirements once a week by preparing two main dishes, each in sufficient quantity to serve four people three times. This will make a total of six complete meals for four people. One meal, say a liver casserole, is served straight away: a second meal, perhaps using lamb chops, will be refrigerated until it can be eaten the next day; which leaves four meals, two of each kind, to be frozen. You will already have a good assortment of meals in the freezer so that you can choose from these for the rest of the week, ensuring that you don't have to eat liver casserole and lamb chops the whole week through!

It is a good notion to give each complete meal a slight variation before freezing – or eating – it. If you make a steak and kidney casserole, you can give that day's meal a pastry crust, and the two meals destined for the freezer can be given trimmings to make them a little different (for example, cobbler topping).

You will already have noticed that six meals are prepared each week, while there are seven days to be dealt with! This isn't an oversight, but it has been assumed that on at least one day out of every week, you will be invited to someone else's home for a meal, or perhaps friends will be coming to you, in which case you will not be feeding them from the family-meal section of the freezer. Your husband might unexpectedly take you out to a restaurant meal or bring home something for dinner!

The dishes you decide to prepare and freeze each week can depend on the foodstuffs that happen to be a particularly good buy at the time. For example, chicken or lamb dishes when they are especially cheap. Many people are often deterred from buying big, deliciously juicy cuts of ham and gammon because they feel that they are going to be eating ham for days on end. Now you can buy and cook it, eat some of the ham straight away and freeze the remainder, made into all kinds of interesting dishes.

Soup is another wonderful family stand-by and nothing you buy, however good, is quite the equivalent of home-made soup. Serve it with hot toast and home-made pâté for a family lunch or supper.

Desserts can be whipped out of the freezer in no time – frozen purées are excellent for emergency desserts, so too is frozen fruit, which can be served with little cupcakes also stored in the freezer.

Individual dinners find their place in this section, for sometimes a member of the family is going to eat a solitary meal – either because he is going to be late or because everyone else has gone out! Use small quantities left over from bulk cooking for single portions, packed in foil plates. Incidentally it may be a little expensive to invest in aluminium foil plates for individual plate dinners, but if they are carefully treated they can be re-used and so minimise the cost.

Rich beef stew with mustard croûtons (page 131)

Pork satay with golden and white rice (page 133)

Green summer soup

1 oz./25 g. butter
1 small bunch spring onions, chopped
3 sticks celery, chopped
8 oz./225 g. leaf spinach
1 Webbs Wonder lettuce, shredded
1 pint/6 dl. chicken stock

Melt the butter and sauté the spring onions and celery until softened but not coloured. Add the spinach, stir well and cook for about 5 minutes. Add the lettuce and stock to the pan and bring to the boil, stirring constantly. Cover and simmer for 10 minutes. Cool, then liquidise or sieve.

To freeze: Pack into three plastic containers. Seal and freeze.

To prepare for serving: Hold the container under running cold water and turn the contents into a large saucepan. Add twice the quantity of chicken stock and reheat, stirring occasionally. Taste and correct seasonings. Or, add one part stock and one part béchamel sauce to give a creamy textured soup. *Makes 3 servings for 4.*

Winter vegetable soup

4 sticks celery, thinly sliced
8 oz./225 g. parsnips, cut into $\frac{1}{4}$-inch/$\frac{1}{2}$-cm. dice
8 oz./225 g. turnips, cut into $\frac{1}{4}$-inch/$\frac{1}{2}$-cm. dice
2 onions, peeled and finely sliced
bouquet garni plus 6 peppercorns
4 pints/2$\frac{1}{4}$ litres chicken stock or 4 stock cubes and water
4 oz./100 g. mushrooms, finely sliced

Put all the vegetables, except the mushrooms, into a saucepan. Tie the bouquet garni and peppercorns in muslin and place in the pan. Add the stock and simmer for 20 minutes. Add the mushrooms and continue to simmer for another 10 minutes.

To freeze: Cool and remove the herbs. Pack into 1$\frac{1}{2}$-pint (1-litre) containers, leaving a $\frac{1}{2}$–1-inch (1–2$\frac{1}{2}$-cm.) headspace.

To prepare for serving: Immerse the container in hot water to remove the contents. Reheat gently in a covered saucepan. Dilute, if liked. Correct the seasoning. *Makes 3 servings for 4.*

Buckling pâté

3 large bucklings
6 oz./175 g. butter, softened
1 clove garlic, crushed
juice of 2 lemons
freshly ground pepper

Put the bucklings in boiling water for 1 minute, so that the skin can be removed easily. Remove the skin and bones. Pound the flesh with a wooden spoon and blend together with the butter. Add the crushed garlic and lemon juice. Season to taste.

To freeze: Put the pâté into three foil dishes, smooth the tops and seal with a double thickness of foil.

To prepare for serving: Thaw at room temperature, for 3–4 hours. Turn out of the container and cut into quarters. Serve with toast. *Makes 3 servings for 4.*

Country pâté

12 oz./350 g. pigs' liver
8 oz./225 g. fat bacon
1 onion, peeled and chopped
1 clove garlic, chopped
2 oz./50 g. butter
salt and pepper to taste
$\frac{1}{2}$ pint/3 dl. well-seasoned white sauce

Fry the liver, bacon, onion and garlic in the butter for about 10 minutes, gently. Remove from the heat and mince or put in a blender. Season with salt and pepper. Combine the liver mixture with the sauce. Turn into a 1$\frac{1}{2}$-pint (1-litre) greased terrine or pie dish, cover with a lid or foil and place in a baking tin of hot water. Bake in a moderate oven (350°F., 180°C., Gas Mark 4) for 1 hour. Cover with foil, stand kitchen weights on top and leave overnight in the refrigerator.

To freeze: Turn out of the dish, cut into slices and stack with pieces of foil between slices. Wrap in double or heavy duty foil. Seal and freeze.

To prepare for serving: Unwrap the required number of slices. Allow about 1 hour for thawing. Serve with hot toast. *Makes 3 servings for 4.*

Chicken liver pâté

3 oz./75 g. butter
12 oz./350 g. chicken livers
salt and pepper to taste
¼ teaspoon powdered mace

Melt 1 oz. (25 g.) of the butter in a frying pan, add the chicken livers and cook gently for 5 minutes. Pass the livers through a fine mincer, melt the remaining butter and add to the chicken livers with the rest of the ingredients.

To freeze: Press into three foil pie dishes and seal with double or heavy duty foil, or store individual portions using the following method: Chill the pâté in the refrigerator, turn out of the foil pie dishes and cut into quarters. Stack the slices with dividing pieces of foil or sheets of polythene tissue paper between each. Wrap in foil and seal.

To prepare for serving: Thaw, covered, at room temperature for approximately 1 hour. *Makes 3 servings for 4.*

Bacon pâté

3 oz./75 g. butter
1 large onion, finely chopped
1½ lb./675 g. cooked bacon, minced
1 tablespoon chopped parsley, blanched
1 tablespoon freshly made mustard
½ teaspoon Worcestershire sauce

Melt the butter in a saucepan and gently fry the onion in it. Combine with the remaining ingredients. Spread the mixture evenly in a 2-lb. (1-kg.) loaf tin, smooth the top and chill in the refrigerator for 2 hours.

To freeze: Turn the pâté out of the loaf tin. Cut into 12 slices and stack with foil dividers between the slices. Wrap in double or heavy duty foil, seal and freeze.

To prepare for serving: Take the required number of slices out of the package. Thaw at room temperature allowing approximately 1 hour. *Makes 3 servings for 4.*

Barbecued meat sauce

4 medium onions, peeled and chopped
3 cloves garlic, crushed
1 head celery, trimmed and chopped
2 oz./50 g. lard or dripping
4 lb./1¾ kg. minced beef
4 teaspoons salt
½ teaspoon pepper
3 tablespoons Worcestershire sauce
¾ pint/4 dl. tomato ketchup

Fry the onion, garlic and celery gently in the fat in a large pan. Gradually add the minced beef and stir until all the meat has browned. Add the salt, pepper, Worcestershire sauce and ketchup. Simmer gently for 20 minutes. Skim off any excess fat.

To freeze: Cool quickly. Spoon the mixture into three 1½-pint (1-litre) containers. Seal, label and freeze.

To prepare for serving: Immerse the container in hot water, to allow the mixture to slip out of the container into a pan for heating. This sauce can be served with all types of pasta or with rice to make a main meal. It can be heated while the rice or pasta is cooking. *Makes 4½ pints (2½ litres).*

Barbecue sauce

2 oz./50 g. butter
2 large onions, peeled and chopped
8 tablespoons Worcestershire sauce
8 tablespoons cider or wine vinegar
2 tablespoons brown table sauce
2 oz./50 g. sugar
2 pints/generous litre water

Melt the butter in a saucepan and sauté the onion. Add the other ingredients and simmer, covered, for 20 minutes. Uncover, then boil rapidly to reduce by one third.

To freeze: Cool. Strain into three ½-pint (3-dl.) containers.

To prepare for serving: Immerse the container in hot water. Turn into a saucepan, simmer gently. *Makes 1½ pints (scant litre).*

Taramasalata

12 oz./350 g. smoked cod's roe
½ pint/3 dl. olive oil
juice of 1 lemon
freshly ground black pepper
1 teaspoon very finely chopped onion
2 tablespoons chopped parsley, blanched

Remove the skin from the roe and discard. Put the roe in a bowl with 5 tablespoons of the oil. Allow to stand for 10 minutes. Pass the mixture through a fine sieve or liquidise in a blender. Add the lemon juice. Beat in the remaining olive oil, 1 tablespoon at a time. Add the remaining ingredients.

To freeze: Pack into individual foil pie dishes. Smooth the tops and seal with double or heavy duty foil. Freeze.

To prepare for serving: Thaw at room temperature allowing approximately 1 hour. Serve on toast as an appetiser or savoury. *Makes 3 servings for 4.*

Tomato sauce

1 medium onion, chopped
2 cloves garlic, crushed with 1 teaspoon salt
1 tablespoon oil
3 tablespoons tomato purée
2 2 lb. 3-oz./992-g. cans tomatoes
2 tablespoons castor sugar
3–4 parsley stalks
1 bay leaf
salt and pepper to taste

Sauté the onion and garlic in the oil until soft. Stir in the tomato purée, canned tomatoes (with the juice) and sugar. Tie the parsley stalks and bay leaf with thread and add to the ingredients in the pan; add the salt and pepper. Bring to the boil, reduce the heat then cover with a lid and simmer gently for 45 minutes. Remove the parsley stalks and bay leaf.

To freeze: Cool rapidly. Pour into suitable containers, allowing a 1-inch (2½-cm.) headspace. Cover and freeze.

To prepare for serving: Reheat in a covered saucepan, over gentle heat. *Makes 4 pints (2¼ litres).*

Curry sauce

3 tablespoons cooking oil
2 large onions, peeled and finely chopped
2 cloves garlic, crushed with 1 teaspoon salt
3 tablespoons tomato purée
2 tablespoons curry powder
2 teaspoons curry paste
1 tablespoon brown sugar
2 tablespoons mango chutney, chopped
1 11-oz./312-g. can tomatoes
1 bay leaf
1 pint/6 dl. stock or stock and gravy, mixed

Heat the oil in a saucepan, add the onions and garlic and cook gently until soft but not brown. Stir in the tomato purée, curry powder and curry paste, cook for 2 minutes. Add the remaining ingredients, cover with a lid and simmer for 45 minutes.

To freeze: Cool and pack into ½-pint (3-dl.) plastic containers, leaving a ½-inch (1-cm.) headspace. Seal and freeze.

To prepare for serving: Immerse in hot water to loosen the contents. Reheat gently in a saucepan. Use for curried eggs, chicken or vegetables. *Makes approximately 1½ pints (scant litre).*

Onion sauce

1 lb./450 g. onions, peeled and chopped
2 tablespoons cornflour
½ pint/3 dl. milk
1 chicken stock cube

Put the onions in a saucepan, cover with water and bring slowly to the boil. Drain and return the onions to the pan, cover with fresh water, bring to the boil and simmer until tender. Drain and reserve ½ pint (3 dl.) of the liquid.

In a mixing bowl, mix the cornflour with 3 tablespoons of the milk, to make a smooth paste.

Combine the remaining milk with the onion water and bring to the boil. Meanwhile, stir in the crumbled stock cube.

Pour the boiling liquid on to the cornflour mixture, stirring all the time, return to the pan, add the onions and continue to cook for 2 minutes, stirring constantly. Cover the surface of the sauce with greaseproof paper and cool.

To freeze: Pour into $\frac{1}{4}$-pint ($1\frac{1}{2}$-dl.) used cream cartons or plastic containers. Seal and freeze.

To prepare for serving: Immerse the container in hot water to loosen the contents. Reheat gently. *Makes $1\frac{1}{2}$ pints (scant litre).*

Bread sauce

$1\frac{1}{2}$ pints/scant litre milk
1 small onion, peeled and finely chopped
1 blade mace
6 white peppercorns
$\frac{1}{2}$ teaspoon salt
6 oz./175 g. breadcrumbs, freshly made
2 oz./50 g. butter

Heat the milk in a saucepan and add the onion, mace and peppercorns. Simmer for 30 minutes. Strain into a clean saucepan, add the salt, breadcrumbs and butter. Simmer gently until the sauce is of a thick, creamy consistency.

To freeze: Cool and pack into used $\frac{1}{4}$-pint ($1\frac{1}{2}$-dl.) cream cartons or plastic containers, leaving a $\frac{1}{2}$-inch (1-cm.) headspace. Seal with a double thickness of foil.

To prepare for serving: Immerse the carton in hot water to loosen the contents. Turn into a small pan and reheat over gentle heat. Add 1 tablespoon cream before serving, if liked. *Makes $1\frac{1}{2}$ pints (scant litre).*

Salmon fish cakes

3 $7\frac{1}{2}$-oz./211-g. cans salmon
$3\frac{1}{2}$ lb./$1\frac{1}{2}$ kg. potatoes, cooked and mashed
2 tablespoons chopped parsley, blanched
salt and pepper to taste
2 beaten eggs and breadcrumbs, for coating

Salmon fish cakes

Drain and flake the salmon, removing any skin and bones. Combine with the potatoes, parsley and seasoning. Divide the mixture in half and roll both pieces into sausage shapes. Cut each into 18 equal-sized pieces. Shape into cakes; dip in the beaten egg and coat with breadcrumbs.

To freeze: Either wrap the fish cakes in a double thickness of foil or pack in an air-tight container, with sheets of polythene tissue paper or foil between the layers. Seal and freeze.

To prepare for serving: Heat fat in a frying pan. Shallow-fry the required number of unthawed fish cakes. *Makes 3 servings for 4.*

American fish pie

2½ lb./generous kg. fresh cod or haddock, cooked
½ pint/3 dl. white sauce, coating consistency
1 oz./25 g. parsley, blanched and chopped
salt and pepper to taste
¼ teaspoon nutmeg
juice of ½ lemon
3 lb./1¼ kg. potatoes, mashed
1 lb./450 g. tomatoes, skinned and sliced
12 oz./350 g. cheese, grated

Remove the skin and bones from the fish, flake without mashing. Add to the sauce with the parsley, seasonings and lemon juice.

To freeze: Arrange alternate layers of mashed potatoes and fish, with a layer of tomatoes and cheese between them, in three foil pie dishes, loaf tins or other suitable containers. Finish with layers of mashed potato. Smooth the tops and seal with a double thickness of foil.

To prepare for serving: Uncover the container and bake unthawed in the centre of a hot oven (400°F., 200°C., Gas Mark 6) for 40–45 minutes. Fork up the potato after the first 10 minutes of cooking time, to give a better appearance. *Makes 3 servings for 4.*

Fish chowder

3 lb./1¼ kg. smoked haddock
bouquet garni
1 thick slice lemon
2 pints/generous litre water
1½ oz./40 g. butter
1 large onion, peeled and finely chopped
2 oz./50 g. cornflour
2 pints/generous litre milk
2 teaspoons tomato purée mixed with juice of ½ lemon
pinch pepper
¼ teaspoon ground mace
6 oz./175 g. Patna rice, boiled

Rinse the haddock and cut into large pieces. Put it into a large saucepan. Place the bouquet garni in the pan with the lemon slice and water. Bring to the boil, then lower the heat and simmer gently until the fish is cooked (approximately 15 minutes). Remove the bouquet garni and lemon slice, drain

off the liquid and set aside. Cool the fish, remove the skin and bones, then flake without mashing. Rinse the saucepan and melt the butter in it, add the onion and cook until soft but not brown. Blend the cornflour with ¼ pint (1½ dl.) of the milk; add to the pan with the remaining milk, reserved fish stock, tomato mixture, pepper and mace. Stir until boiling then reduce the heat and continue to cook for a further 2–3 minutes.

To freeze: Cover the liquid in the pan with a buttered paper and allow to cool. Mix in the flaked fish and rice. Pack in polythene bags, plastic or other suitable containers, leaving a ½-inch (1-cm.) headspace. Freeze.

To prepare for serving: Thaw, overnight, in the refrigerator. Reheat gently in a saucepan. Sprinkle each serving with chopped parsley. *Makes 3 servings for 4.*

Beef goulash

1½ oz./40 g. lard or dripping
3½ lb./1½ kg. stewing steak, trimmed and cut into
　　1-inch/2½-cm. cubes
1 clove garlic
1 teaspoon caraway seeds
1½ lb. onions, peeled and sliced
1 tablespoon paprika pepper
1 11-oz./312-g. can tomatoes
1 teaspoon sugar
salt and pepper to taste
¼ pint/1½ dl. stock or water

Heat the lard in a large saucepan. Brown the meat, a third at a time, over a brisk heat. Crush the garlic with the caraway seeds and add to the pan with the onions, lower the heat and cook until the onions are soft. Add the remaining ingredients, cover with a lid and simmer very gently for 1¼ hours.

To freeze: Cool rapidly and pack into foil or plastic containers, or partially freeze in a covered, foil-lined meat tin, then cut into squares and wrap each portion in a double thickness of foil. Return to the freezer.

To prepare for serving: Thaw at room temperature for about 4 hours. Reheat in a saucepan. Add

$\frac{1}{4}$ oz. (10 g.) beurre manié (see page 130) and cook for a further 2 minutes. Remove from the heat and add 3 tablespoons yogurt. *Makes 3 servings for 4.*

Steak, kidney and mushrooms

3 lb./1$\frac{1}{4}$ kg. stewing steak, cut into 1-inch/2$\frac{1}{2}$-cm. cubes

12 oz./350 g. ox kidney, cored and cubed

2 oz./50 g. seasoned flour

2 oz./50 g. lard or dripping

2 large onions, peeled and sliced

12 oz./350 g. mushrooms, sliced

1$\frac{1}{2}$ pints/scant litre chicken stock or
   2 stock cubes and water

Toss the steak and kidney in the flour. Heat the fat in a frying pan. Add the onions, fry gently for 5 minutes, transfer to a casserole dish. Put the steak and kidney into the pan and fry briskly to brown. Put into the casserole dish, add the mushrooms and stock. Cover with a lid and cook in a moderate oven (350°F., 180°C., Gas Mark 4) for 1$\frac{1}{2}$ hours.

To freeze: Cool rapidly and pack in three foil, plastic or other suitable containers, cover and freeze.

To prepare for serving: Reheat without thawing in a moderate oven, or gently in a pan. *Makes 3 servings for 4.*

Farmhouse brisket

8 oz./225 g. streaky bacon, diced

4 lb./1$\frac{3}{4}$ kg. brisket, boned and rolled

1 lb./450 g. onions, sliced

1$\frac{1}{2}$ lb./675 g. carrots, sliced

3 sticks celery, cut into $\frac{1}{2}$-inch/1-cm. pieces

1$\frac{1}{2}$ oz./40 g. plain flour

1$\frac{1}{2}$ pints/scant litre brown ale

1 tablespoon brown sugar

1 teaspoon salt

pinch pepper

1 bay leaf

Fry the bacon gently, to render down the fat. Transfer to a casserole dish. Fry the meat, briskly, in the fat, turning to brown on all sides. Put

the browned meat in the casserole with the bacon.

Lower the heat under the frying pan, add the onions and fry gently until soft but not brown. Add the carrots and celery and cook for a further 5 minutes. Stir in the flour and cook for another 2 minutes.

Spoon the vegetables around the meat and add the remaining ingredients. Cover with a lid and cook in the centre of a warm oven (325°F., 170°C., Gas Mark 3) for 2–2$\frac{1}{2}$ hours.

To freeze: Cool the meat and cut into slices. Pack flat in three suitable foil or plastic containers, cover closely and freeze.

To prepare for serving: Immerse the container in hot water to loosen the contents, transfer to a saucepan and reheat gently. *Makes 3 servings for 4.*

Braised steak with mushrooms

Braised steak with mushrooms

2 oz./50 g. dripping or lard

3 lb./1$\frac{1}{4}$ kg. braising steak

1$\frac{1}{2}$ lb./675 g. onions, peeled and sliced

1$\frac{1}{2}$ lb./675 g. carrots, peeled and sliced

1 head celery, chopped

6 oz./175 g. mushrooms, sliced

$\frac{1}{2}$ bottle red cooking wine

1$\frac{1}{2}$ pints/scant litre beef stock or 2 stock cubes and water

salt and pepper to taste

Heat the dripping in a large pan, cut the meat into $\frac{1}{2}$-inch (1-cm.) slices and fry briskly to seal on both sides. Remove and keep hot. Add the onions, carrots, celery and mushrooms to the pan and cook for 2 minutes. Put the vegetables and meat into the casserole. Add the wine, stock and seasoning. Cover and cook in a moderately hot oven (375°F., 190°C., Gas Mark 5) for 1½ hours.

To freeze: Cool rapidly. Pack in three foil or plastic containers or in polythene bags. Seal and freeze.

To prepare for serving: Remove from freezer 2 hours before use. Heat in a moderately hot oven (375°F., 190°C., Gas Mark 5) for 35–40 minutes. *Makes 3 servings for 4.*

Portuguese cobbler

Portuguese cobbler

2 oz./50 g. dripping or lard

2 large onions, peeled and chopped

2½ lb./generous kg. pie veal, cut into 2-inch/5-cm. cubes

2 oz./50 g. flour

2 15-oz./425-g. cans tomato soup

1 lb./450 g. carrots, peeled and sliced

salt and pepper

2 green peppers, de-seeded and diced

6 oz./175 g. mushrooms, sliced

For the topping:

12 oz./350 g. plain flour

1½ teaspoons baking powder

1 teaspoon salt

4 oz./100 g. margarine or lard

¼ pint plus 2–3 tablespoons/1½ dl. milk

Melt the fat in a 4-pint (2-litre) flameproof dish, fry the onions and meat until lightly browned, remove them from the dish. Stir the flour into the fat, add the soup and bring to the boil. Return the onions and meat to the dish, add the carrots

and season to taste. Cook in a slow oven (300°F., 150°C., Gas Mark 2) for 2 hours. Add the green peppers and mushrooms 20 minutes before the end of the cooking time. Remove from the oven and turn the thermostat up to 425°F., 220°C., Gas Mark 7. Sieve the flour, baking powder and salt into a mixing bowl. Rub in the fat and add sufficient milk to make a soft dough. Cut into 1½–2-inch (4–5-cm.) circles, place on a greased baking sheet and brush with milk. Bake in the top of the pre-heated oven for 10–12 minutes. Cool on a wire rack.

To freeze: Cool the meat mixture and pack in three foil pie dishes. Seal with double thicknesses of foil. Pack cobblers separately in polythene bags. Seal tightly and freeze.

To prepare for serving: Place unthawed meat in a pre-heated oven (325°F., 170°C., Gas Mark 3) for about 1 hour. 10 minutes before the end of cooking time, remove the foil and arrange the cobblers to cover the surface of the meat. *Makes 3 servings for 4.*

Osso buco

6 lb./2¾ kg. shin of veal, chopped in 2-inch/5-cm. pieces

2 oz./50 g. seasoned flour

4 oz./100 g. butter or margarine

2 carrots, peeled and cut into thick slices

1 onion, peeled and sliced

1 clove garlic

1 teaspoon salt

1 2¼-oz./64-g. can tomato purée

¾ pint/4 dl. chicken stock or 1 stock cube and water

pinch pepper

1 bay leaf

1 teaspoon sugar

Coat the veal with seasoned flour. Melt the butter in a large frying pan and put the carrot and onion into the pan. Peel the garlic, crush with the salt and add it to the vegetables in the pan. Fry gently for 5 minutes. Transfer the vegetables to a large saucepan. Put the veal into the frying pan and brown on all sides, over strong heat. Transfer the meat to the saucepan and add the remaining ingredients. Cover with a lid and simmer gently for 2 hours. Strain the sauce and reserve.

To freeze: Cool, discard the vegetables and bay leaf. Remove the meat from the bones. Pack the veal into three plastic, foil or other suitable containers. Divide the sauce between the three servings. Seal and freeze.

To prepare for serving: Immerse the container in hot water to loosen the contents. Reheat in a covered saucepan over gentle heat. *Makes 3 servings for 4.*

Country braise

Country braise

2 oz./50 g. lard or dripping

2 large onions, peeled and sliced

1 lb./450 g. carrots, peeled and sliced

1 large cooking apple, peeled, cored and sliced

3 lb./1¼ kg. stewing veal, cut into 1-inch/2½-cm. cubes

2 oz./50 g. seasoned flour

¾ pint/4 dl. chicken stock or 1 stock cube and water

3 tablespoons tomato ketchup

4 oz./100 g. prunes, soaked for 2 hours and stoned

2 oz./50 g. almonds, blanched and chopped

Heat the lard in a large saucepan, add the onions, carrots and apples and fry until lightly browned; remove from the pan. Toss the veal in the flour and fry gently for 5 minutes. Return the vegetables to the saucepan, add the stock and tomato ketchup, bring to the boil then simmer very gently for 1 hour.

To freeze: Cool, add the prunes and almonds. Pack in three plastic containers or polythene bags, seal and freeze.

To prepare for serving: Thaw at room temperature, allowing approximately 4 hours. Turn into a saucepan, bring to the boil, cover with a lid and simmer for 30 minutes. *Makes 3 servings for 4.*

Devilled rabbit

12 rabbit pieces

2 tablespoons malt vinegar

2 tablespoons mild mustard, freshly made

1 tablespoon curry powder

5 tablespoons corn oil

12 oz./350 g. pickled pork, diced

2 onions, peeled and chopped

seasoned flour, for coating

1½ pints/scant litre chicken stock or
    2 stock cubes and water

4 tablespoons cornflour

6 tablespoons milk

salt and pepper to taste

1 teaspoon white vinegar

Put the rabbit pieces into a bowl of cold water, add 2 tablespoons vinegar and leave to soak for about 1 hour. Drain and pat dry. Mix the mustard and curry powder. Spread over the rabbit joints. Leave for another 30–40 minutes for the flavour to penetrate. Heat the oil in a heavy pan. Add the pork and onion. Cook the onion until transparent but not brown. Dip the rabbit pieces into the flour; add them to the pan and brown lightly on all sides. Pour the stock over and stir until boiling. Cover with a lid and simmer gently for 1 hour. Blend the cornflour with the milk and add to the pan. Remove from the heat, check the seasoning and add the white vinegar. Cool.

To freeze: Pack into three foil-lined serving dishes. Cover with foil and partially freeze. Remove from the serving dishes, when sufficiently solid to do so, and over-wrap with double or heavy duty foil. Seal and freeze.

To prepare for serving: Unwrap and return to the original dish. Reheat in a hot oven (400°F., 200°C., Gas Mark 6). *Makes 3 servings for 4.*

Liver and onion casserole

3 lb./1¼ kg. ox liver, sliced and trimmed
2 oz./50 g. seasoned flour
3 oz./75 g. dripping or lard
2 lb./900 g. onions, peeled and sliced
¾ pint/4 dl. beef stock or 1 stock cube and water
½ teaspoon mixed herbs

Soak the liver in cold, salted water for 30 minutes. Dry on absorbent paper then coat in the seasoned flour. Melt the fat in a heavy saucepan or flame-proof casserole and cook the onion gently until soft but not brown. Remove the onion and set aside. Fry the liver briskly in the remaining fat to seal on both sides. Return the onion to the pan and add any remaining seasoned flour. Stir well, then add the stock and herbs. Cover the casserole with a lid and cook in a warm oven (325°F., 170°C., Gas Mark 3) for 1½ hours.

To freeze: Cool rapidly. Spoon into three plastic or foil containers. Seal and freeze.

To prepare for serving: Thaw overnight in the refrigerator. Reheat gently, in a covered pan. *Makes 3 servings for 4.*

Cutlet turnovers

12 large lamb cutlets, boned
1½ lb./675 g. puff pastry
½ pint/3 dl. onion sauce (see page 83)

Trim the fat from the meat and roll tightly. Secure each with a cocktail stick. Grill and set aside to cool. Remove the cocktail sticks.

For ease of handling, cut the pastry into three equal pieces. Roll each piece into a 16-inch (40-cm.) square, trim the edges and cut into quarters. Spoon a little sauce into the centre of each square and top with a lamb cutlet. Damp two adjoining edges of each square and fold over to make a triangle. Seal and pinch the edges. Place in the refrigerator to chill the pastry thoroughly.

To freeze: Pack upright, in a suitable polythene container, with foil or polythene tissue dividing papers between each. Seal and freeze.

To prepare for serving: Place, unthawed, on a wet baking sheet. Brush with milk, cook in a very hot oven (450°F., 230°C., Gas Mark 8) for 20 minutes. *Makes 12.*

Moussaka

Moussaka

2 oz./50 g. lard or dripping
1 large onion, peeled and finely chopped
3 lb./1¼ kg. raw shoulder lamb, minced
salt and pepper to taste
3 tablespoons tomato purée
1 tablespoon sugar
3 large aubergines, sliced and sprinkled with salt
1 clove garlic, crushed
1 11-oz./312-g. can tomatoes
¾ pint/4 dl. cheese sauce
3 oz./75 g. cheese, grated

Heat 1 oz. (25 g.) of the lard or dripping in a saucepan. Add the onion and allow to colour. Add the meat and stir over a brisk heat to brown. Season with salt and pepper, add the tomato purée and sugar. Set aside. Wash the aubergines under cold, running water. Dry on kitchen paper. Heat the remaining fat in the pan, add 2 aubergines to the pan and cook for 5–7 minutes, add the garlic and tomatoes. Cover the pan and continue to cook for a further 5 minutes.

To freeze: Divide the meat mixture between three foil dishes. Cover with the aubergine and tomato mixture. Top with the cheese sauce and remaining aubergine slices. Sprinkle 1 oz. (25 g.) cheese on each dish. Seal with double thicknesses of foil and freeze.

To prepare for serving: Uncover, place in a hot oven (400°F., 200°C., Gas Mark 6) for 40 minutes. *Makes 3 servings for 4.*

Potted hough

1½ lb./675 g. shin of beef

2–2½ lb./about 1 kg. knuckle of veal, chopped by the
  butcher

1½ teaspoons salt

6 peppercorns

1 bay leaf

Put the beef, veal and salt into a large saucepan,
add enough water to cover and bring to the boil.
Simmer until the meat is tender (approximately
3 hours). Strain the liquid into a clean saucepan.
Cool the meat, cut into small pieces and divide
between three 1-pint (6-dl.) foil pudding basins.

Put the bones into the stock with the peppercorns and bay leaf. Boil rapidly, without a lid, to
reduce to ½ pint (3 dl.). Strain the stock on to the
meat and leave to set.

To freeze: Cover the basins with double or heavy
duty foil, seal and freeze.

To prepare for serving: Thaw overnight in the
refrigerator. Serve with salad.   *Makes 3 servings
for 4.*

Terrine of game

2 oz./50 g. butter

1 pheasant, plucked and drawn

2 partridges, plucked and drawn

12 oz./350 g. shin of veal, chopped into 2-inch/5-cm.
  pieces

3 pints/1¾ litres stock or water

3 cloves

10 white peppercorns

2 teaspoons salt

¼ pint/1½ dl. dry white wine

Melt the butter in a large, heavy saucepan. Brown
the birds over a brisk heat. Put all the birds back
into the pan with the veal, stock, cloves, peppercorns and salt. Bring to the boil; lower the heat,
cover and simmer for 1½–2 hours. Remove the
birds and the veal from the pan, cool and cut into
pieces. Discard the bones. Strain the stock into a
clean saucepan, add the wine and boil rapidly to
reduce by two-thirds. Meanwhile, arrange the
game in three foil-lined terrines or small pie
dishes. Skim the fat from the stock and pour it

over the game. (Older birds may be used for this
dish.)

To freeze: Cool rapidly, until set. Cover with foil
and partially freeze. Remove from the terrine and
over-wrap with a double or heavy duty foil.
Return to the freezer.

To prepare for serving: Thaw overnight in the
refrigerator. Serve in the original dish.   *Makes 3
servings for 4.*

Beef galantine

2 lb./900 g. rump steak

12 oz./350 g. cooked ham

1 large onion, peeled and quartered

1½ lb./675 g. tomatoes, skinned

6 oz./175 g. white breadcrumbs

2 eggs, lightly beaten

salt and pepper to taste

Cut away any fat or gristle from the steak and ham.
Mince the steak, ham and onion finely, by passing
it through the mincer two or three times. Blend
or sieve the tomatoes and add to the meat mixture
with the remaining ingredients. Mix thoroughly
to combine.

Divide the mixture into three equal portions.
Press each into a 1-pint (6-dl.) foil basin. Cover
with aluminium foil and steam gently for 3 hours.

Cool in the refrigerator or larder with a kitchen
weight on each basin.

To freeze: When quite cold, re-cover the dishes
with foil and seal before freezing.

To prepare for serving: Thaw overnight in the
refrigerator or at room temperature for 4 hours.
Makes 3 servings for 4.

Cornish pasties

1½ lb./675 g. shortcrust pastry

For the filling:

1¼ lb./550 g. lean beefsteak

2 medium potatoes, diced

2 medium onions, peeled and finely chopped

2 tablespoons chopped, blanched parsley

salt and pepper to taste

1 tablespoon stock or water

Mince the meat finely and add to the other filling ingredients. Divide the pastry into 12 portions and roll each piece into a 5-inch (13-cm.) circle. Put a spoonful of the filling into the centre of each piece of pastry, damp the edges, fold in half and seal. Flute the edges by pinching between the thumb and first two fingers. Stand the pastries upright on a greased baking sheet, brush with egg and milk and set aside in a cool place for 30 minutes, to allow the pastry to relax. Bake in the centre of a moderately hot oven (375°F., 190°C., Gas Mark 5) for 35 minutes.

To freeze: Allow to become completely cold. Pack in plastic, aluminium foil or other suitable containers, seal and freeze.

To prepare for serving: Allow to thaw at room temperature for approximately 4 hours. Reheat in a cool oven. *Makes 3 servings for 4.*

Devonshire pasties

1 lb./450 g. shortcrust pastry
For the filling:
1 lb./450 g. raw beef, minced
12 oz./350 g. mushrooms, chopped
1 11-oz./312-g. can sweetcorn with peppers
salt and pepper
beaten egg

Mix the minced meat, mushrooms and drained sweetcorn and season with salt and pepper. Roll out the pastry to about ⅛ inch (3 mm.) thick and cut into circles around a saucer. Divide the filling evenly between the pasties, damp the edges with water, fold over and seal. Brush with beaten egg and make a couple of slashes on top of each pasty. Bake in a hot oven (425°F., 220°C., Gas Mark 7) for about 30 minutes, when the pastry should be crisp and golden.

To freeze: Allow to become completely cold. Pack into a suitable container, cover closely and freeze.

To prepare for serving: Reheat frozen pasties in a moderate oven (350°F., 180°C., Gas Mark 4) for approximately 20 minutes. *Makes 3 servings for 4.*

Chicken special

2 medium-sized chickens
3 pints/1¾ litres water
1 lb./450 g. streaky bacon
1 lb./450 g. button mushrooms
3 oz./75 g. butter
3 oz./75 g. flour
salt and pepper to taste

Simmer the chickens in the water until tender. Set aside until cool. Reserve the stock for the sauce. Chop the bacon into 2-inch (5-cm.) pieces after removing the rind. Cut the mushrooms in half. Fry the bacon until crisp and remove from the pan. Cook the mushrooms gently in the bacon fat. Melt the butter in a clean saucepan, add the flour and cook for 1 minute. Season to taste, remove from the heat and gradually stir in 1½ pints (scant litre) stock. Return to the heat and bring to the boil, stirring. Remove all the flesh from the chickens and cut into bite-size pieces; add to the sauce with the bacon and mushrooms.

To freeze: Cool rapidly, pack in foil, plastic or other suitable containers. Seal tightly and freeze.

To prepare for serving: Immerse the container in hot water to loosen the contents. Turn into a saucepan and reheat gently. Add a small packet of frozen peas and continue to cook for a further 7 minutes. *Makes 3 servings for 4.*

Creamed pimento chicken pie

6 oz./175 g. butter
5 oz./150 g. flour
salt and pepper to taste
2 pints/generous litre chicken stock or
    2 stock cubes and water
1 pint/6 dl. milk
2½ lb./generous kg. cooked chicken, chopped
1 small can pimento
For the pastry crust:
1 lb./450 g. flaky pastry

Melt the butter over low heat, blend in the flour smoothly and add the seasonings. Gradually stir in the chicken stock and the milk. Bring to the boil, stirring constantly, allow to boil for 1 minute. Stir in the chicken, and the finely chopped pimento together with the liquid from the can. Cool and pack into three foil pie dishes.

To make the crust, divide the pastry into three, roll out thinly and use to cover the pies. Damp the edges and decorate without breaking the seal. Pierce a hole in the centre and bake in a hot oven (425°F., 220°C., Gas Mark 7) for 15 minutes. If necessary, reduce the heat slightly and continue cooking for another 5–10 minutes or until golden brown.

To freeze: Wrap each pie separately in double or heavy duty foil. Seal tightly and freeze.

To prepare for serving: Thaw in the refrigerator and reheat in a moderately hot oven (375°F., 190°C., Gas Mark 5) for about 1 hour.

Note: The pies can be frozen with the pastry unbaked, after sealing the edges of the crusts. Do not pierce a hole in the pastry. Place unbaked in the freezer to harden the pastry, then wrap in foil. Freeze. *Makes 3 servings for 4.*

Scalloped turkey with rice

8 oz./225 g. Patna rice

3 oz./75 g. butter or margarine

1 lb./450 g. mushrooms

3 oz./75 g. flour

salt and pepper to taste

$\frac{3}{4}$ pint/4 dl. hot milk

1 pint/6 dl. hot chicken stock

$2\frac{1}{2}$ lb./generous kg. cooked turkey, chopped

1 medium green pepper, de-seeded

Cook the rice in plenty of boiling salted water until just tender. Drain and rinse in a colander, using fresh, hot water. Melt 1 oz. (25 g.) of the butter and fry the mushrooms for 1 minute only. Remove from the pan, add the rest of the butter and make a thick sauce with the flour, seasoning, milk and stock. Stir constantly over moderate heat until the sauce is smooth and thick.

To freeze: Pack in foil dishes or other suitable containers, with layers of rice, mushrooms and turkey. Pour one-third of the sauce into each container. Chop the pepper finely and sprinkle evenly over the three containers. Cool quickly. Cover closely with foil.

To prepare for serving: Put into the top of a moderately hot oven (375°F., 190°C., Gas Mark 5), turning frequently to heat right through. *Makes 3 servings for 4.*

Cider gammon with apples

Cider gammon with apples

12 slices gammon

black pepper to taste

6 oz./175 g. butter

$1\frac{1}{2}$ pints/scant litre cider

12 oz./350 g. button onions

8 sticks celery

9 tomatoes, quartered

12 oz./350 g. mushrooms

6 dessert apples, cored and sliced

3 oz./75 g. flour

Place the gammon in a casserole, sprinkle in the pepper. Add half the butter and 2 tablespoons of the cider. Cover with a lid and bake in a moderately hot oven (375°F., 190°C., Gas Mark 5) for 15–20 minutes. Meanwhile melt the remaining butter and fry all the vegetables and the apples in it. Add the flour and cook, stirring, for a few minutes. Gradually pour in the cider, and stir until smooth. Pour over the gammon, re-cover the casserole and continue cooking in the oven for a further 20 minutes.

To freeze: Cool rapidly, put into three suitable containers, cover closely and seal before freezing.

To prepare for serving: Thaw at room temperature for about 6 hours. Reheat in a covered casserole in a moderately hot oven. Stir in $\frac{1}{4}$ pint ($1\frac{1}{2}$ dl.) single cream just before serving. *Makes 3 servings for 4.*

Steak and kidney pudding

For the pastry:

1 lb./450 g. self-raising flour

2 teaspoons salt

pinch pepper

8 oz./225 g. shredded beef suet

cold water to mix

For the filling:

2 lb./900 g. stewing steak, cut in 1-inch/2½-cm. cubes

8 oz./225 g. ox kidney, cored and cut in pieces

1 oz./25 g. seasoned flour

6 tablespoons water

Sift together the flour and salt. Add the suet and enough water to make a firm dough. Cut off one-third of the pastry and divide the remaining pastry into three. Use these pieces to line three 1½-pint (1-litre) greased foil pudding basins.

Toss the meat in the seasoned flour and divide between the basins. Pour 2 tablespoons water into each basin. Damp the edges of the pastry and top the basins with the remaining pastry, rolled into circles. Cover with greased foil and steam for 3 hours.

To freeze: Cool quickly. Put into a polythene bag. Seal and freeze.

To prepare for serving: Remove the polythene bag and thaw at room temperature for 6–8 hours. Steam for 1½ hours. *Makes 3 servings for 4.*

Variations

Rabbit and bacon pudding: Omit the steak and kidney and use 2 lb. (900 g.) jointed rabbit, blanched, 1 lb. (450 g.) lean collar bacon, soaked and cut in cubes, 2 large onions, sliced, 4 oz. (100 g.) mushrooms, chopped, 2 cooking apples, peeled, cored and chopped and 1 teaspoon sage.

Veal, gammon and mushroom pudding: Omit the steak and kidney and use 2 lb. (900 g.) pie veal, cut in 1-inch (2½-cm.) cubes, 12 oz. (350 g.) lean gammon, cut into ½-inch (1-cm.) pieces, 8 oz. (225 g.) mushrooms, chopped, grated rind of 1 lemon and ¼ teaspoon dried thyme.

Rhubarb and orange pudding: Omit the steak, kidney, flour and water and use 2 lb. (900 g.) rhubarb, cut in 1-inch (2½-cm.) pieces, the grated rind and chopped segments of 2 large oranges and 6–8 oz. (175–225 g.) sugar. Steam for 2 hours only before freezing.

Gooseberry pudding: Omit the steak, kidney, flour and water and use 2 lb. (900 g.) gooseberries, topped and tailed, and 6–8 oz. (175–225 g.) sugar. Steam for 2 hours only before freezing.

Plum and apple pudding: Omit the steak, kidney, flour and water, and use 2 lb. (900 g.) plums, halved and stoned, 2 cooking apples, peeled, cored and sliced and 8 oz. (225 g.) sugar. Steam for 2 hours only before freezing.

Apricot oat crumble

1 lb./450 g. dried apricots, soaked and stewed

2 oz./50 g. plain flour

4 oz./100 g. butter, melted

6 oz./175 g. demerara sugar

8 oz./225 g. rolled oats

Divide the fruit into three portions and spread into the bases of three 1-pint (6-dl.) foil dishes. Mix the flour, butter and sugar into the rolled oats and scatter over the top of the fruit.

To freeze: Seal with a double thickness of foil and freeze.

To prepare for serving: Remove the cover and sprinkle the surface of the crumble with sugar. Bake in a moderate oven (350°F., 180°C., Gas Mark 4), without thawing, for 40 minutes. *Makes 3 servings for 4.*

Traditional trifle

8 individual sponge cakes, crumbled

6 tablespoons sherry

6 tablespoons raspberry jam

12 macaroons

2 oz./50 g. flaked almonds

¾ pint/4 dl. cold custard

Divide the cake crumbs and press into the bases of 12 individual foil pie plates. Sprinkle with sherry and top with raspberry jam. Sprinkle with crumbled macaroons and flaked almonds, and top with custard.

To freeze: Seal the dishes with foil and freeze.

To prepare for serving: Thaw, covered, at room temperature for about 2 hours. Decorate with whipped cream. *Makes 3 servings for 4.*

Steamed syrup pudding

3 tablespoons golden syrup
12 oz./350 g. self-raising flour
4 teaspoons baking powder
1 teaspoon mixed spice
½ teaspoon salt
8 oz./225 g. luxury margarine
8 oz./225 g. castor sugar
4 large eggs
3 tablespoons milk

Put a tablespoon of syrup into each of three 1-pint (6-dl.) foil pudding basins. Sieve the flour, baking powder, mixed spice and salt into a mixing bowl. Add the remaining ingredients, beat for 2 minutes or until all the ingredients are blended. Spoon equal quantities of the mixture into the pudding basins. Seal the tops of the basins with foil and steam for 1½ hours.

To freeze: Cool, re-seal with double or heavy duty foil.

To prepare for serving: Steam the pudding for approximately 30 minutes, or until hot. *Makes 3 servings for 4.*

Perth pudding and honey sauce

6 oz./175 g. butter
6 oz./175 g. castor sugar
3 eggs
9 oz./250 g. self-raising flour
3 oz./75 g. currants
4½ oz./125 g. sultanas
about 3 tablespoons milk

Cream together the butter and sugar until light and fluffy. Lightly beat the eggs and gradually add to the creamed mixture, beating well. Add a little flour with the egg if the mixture shows signs of curdling. Fold in the remaining flour and the dried fruit. Well grease three 1-pint (6-dl.) pudding basins and two-thirds fill with the mixture. Cover the basins with well-greased double

Perth pudding and honey sauce

or heavy duty foil and crimp tightly under the edges of the basins to seal. Ensure that the water in the steamer or saucepan is boiling, put in the puddings and steam for 1–1½ hours. Remove from the steamer and allow to cool completely.

To freeze: Check seal of foil all around edges. Label and freeze.

To prepare for serving: Get a steamer or saucepan of water boiling, place the frozen pudding in the prepared pan and steam for 40 minutes. To make the sauce, melt together 4 oz. (100 g.) butter and 4 tablespoons clear honey and serve warm with the pudding. *Makes 3 servings for 4.*

Peel, halve and core the pears and place at once in warm syrup.

Arrange the pears carefully in rigid containers, leaving a good headspace, and fill with sufficient syrup to cover.

Place crumpled foil over the pears to keep them submerged.

Pears in spiced honey syrup are delicious served with whipped cream and cinnamon.

Pears in spiced honey syrup

6 oz./175 g. clear honey
1½ pints/scant litre water
3 oz./75 g. crystallised ginger, sliced
good pinch cinnamon
6 tablespoons lemon juice
3 lb./1¼ kg. dessert pears, not quite ripe

Use a wide shallow pan which will allow the pears to lie in a single layer or cook the pears a few at a time. Put the honey and water in the pan and stir over a gentle heat until the honey dissolves. Add the sliced ginger, cinnamon and lemon juice. Peel and halve the pears. Remove the cores and put the fruit immediately into the syrup. The pears must be kept covered with syrup throughout cooking and freezing to avoid discoloration. Poach them gently until they are tender.

To freeze: Stand the pan in cold water to cool them quickly. Arrange pears carefully in plastic containers leaving a ¾-inch (1½-cm.) headspace, cover with syrup, then place crumpled foil on top of the fruit to keep submerged. Cover, freeze.

To prepare for serving: Thaw at room temperature and serve while the fruit is still chilled. Decorate with piped whipped cream and a sprinkling of cinnamon. *Makes 3 servings for 4.*

Raisin ice cream

1 14-oz./396-g. and 1 6-oz./170-g. can evaporated milk, chilled
3 oz./75 g. castor sugar
juice of 2 lemons
6 oz./175 g. seedless raisins
½ teaspoon vanilla essence
1 tablespoon gelatine
3 tablespoons warm water

Whisk all the evaporated milk until thick. Beat in the sugar, lemon juice, raisins and vanilla essence. Dissolve the gelatine in water in a small saucepan. Heat gently, but do not allow to boil.

Quickly stir into the evaporated milk and whisk until on the point of setting. Pour into a plastic container, cover with a lid and freeze for 2 hours. Whisk the mixture to break down the ice particles, cover and freeze again. *Makes 3 servings for 4.*

Butterscotch sauce

6 tablespoons golden syrup
2 oz./50 g. brown sugar
1 oz./25 g. butter
¾ pint/4 dl. water
1 tablespoon custard powder, mixed with
    juice of 1½ lemons

Melt the syrup, brown sugar and butter in a heavy saucepan, heat gently to combine. Remove from the heat and add the water, pour over the custard mixture and return to the saucepan. Stir over gentle heat until thickened.

To freeze: Cool and pack into boiling bags or plastic containers. Seal and freeze.

To prepare for serving: Immerse the bag in boiling water for 10 minutes. Immerse plastic containers in hot water for 15 minutes. Serve hot with ice cream or other desserts. *Makes ¾ pint (4 dl.).*

Sauces for ice cream

Raspberry sauce

1 lb./450 g. raspberry jam
juice of 2 lemons
½ pint/3 dl. water
2 tablespoons arrowroot

Combine the jam, lemon juice and water in a saucepan. Blend the arrowroot with some of the liquid, bring the remainder to the boil and add the blended arrowroot. Cook for a further 2 minutes stirring constantly. Strain to remove the pips.

To freeze: Cool and pack into boiling bags or plastic containers. Seal and freeze.

To prepare for serving: Immerse the bag in boiling water for 10 minutes. Immerse plastic containers in hot water for 15 minutes. Serve hot with ice cream or other desserts. *Makes 1 pint (6 dl.).*

Dark chocolate sauce

1 lb./450 g. plain chocolate
¼ pint/1½ dl. clear honey
¼ pint/1½ dl. lemon juice
4 tablespoons cornflour
1 pint/6 dl. water
8 oz./225 g. unsalted butter

Place all the ingredients in a saucepan and heat gently until the chocolate has melted. Stir briskly until the mixture comes to the boil and is smooth and thick. Remove from the heat and cool, stirring frequently.

To freeze: Cool and pack into boiling bags or plastic containers. Seal and freeze.

To prepare for serving: Immerse the bag in boiling water for 10 minutes. Immerse plastic containers in hot water for 15 minutes. Serve hot with ice cream or other desserts. *Makes 2 pints (generous litre).*

New Zealand guard of honour (page 134)

Frozen Christmas pudding (page 139) and Plum cake ring (page 120)

White bread

Dry mix:

1 lb./450 g. plain flour

2 teaspoons salt

½ oz./15 g. lard

Yeast liquid:

½ oz./15 g. fresh yeast and ½ pint/3 dl. water, or

    2 teaspoons dried yeast, 1 teaspoon sugar and

    ½ pint/3 dl. tepid water (110°F., 43°C.)

Mix the fresh yeast with the water, or dissolve the sugar in the water and sprinkle over the dried yeast. Leave until frothy, about 10 minutes. Sift the flour with the salt and rub in the lard. Mix the dry ingredients with the yeast liquid using a wooden fork or spoon, then work with one hand to a firm dough, adding extra flour, if needed, until the dough leaves the sides of the bowl clean. Turn on to a lightly floured board and knead thoroughly to stretch and develop the dough. To do this fold the dough towards you, then push down and away with the palm of the hand. Repeat for about 10 minutes until the dough feels firm and elastic and no longer sticky. After kneading, shape the dough into a ball. Place in a large, lightly greased polythene bag.

To freeze unrisen dough: Seal the bag tightly and place in the freezer. If the dough might rise slightly before freezing, leave a 1-inch (2½-cm.) space above the dough.

To freeze risen dough: Loosely tie the bag, leaving room for the dough to rise. Allow to rise until double in size:

Quick rise: 45–60 minutes in a warm place.

Slower rise: 2 hours at average room temperature.

Overnight rise: up to 12 hours in a cold larder or refrigerator.

Risen dough springs back when pressed with a lightly floured finger. Turn the risen dough on to a lightly floured board and flatten firmly with the knuckles to knock out air bubbles, then knead until firm. Place in a lightly greased polythene bag, tightly seal and place in the freezer.

To prepare unrisen dough for use: Thaw for about 5–6 hours at room temperature, then leave to rise and knock back, as above.

To prepare risen dough for use: Thaw for about 5–6 hours at room temperature.

To shape the dough: For tin loaves, divide the dough into the quantity required (i.e., leave whole for a large loaf, or divide into two for two small loaves) and shape the loaves as follows. Flatten the dough with the knuckles to an oblong the width of the tin. Fold in three and turn over so that the seam is underneath. Smooth over the top and tuck in the ends. The loaves are now ready to rise again (this is known as proving). Put the tins inside a greased polythene bag and leave until the dough rises to the tops of the tins. This takes 30–40 minutes at room temperature, or leave in a refrigerator if a slow rising is more convenient. Remove from the polythene bag and bake the loaves in the centre of a very hot oven (450°F., 230°C., Gas Mark 8) for 30–40 minutes or until the loaves shrink slightly from the sides of the tins and sound hollow when tapped on the base; cool on a wire tray.

Chapter Eight

Chain cooking

Planning ahead can save you a great deal of time and effort, as well as money. We've seen how this applies to cooking main dishes in large quantities, so that you can serve one meal at once, and freeze the rest. But if the freezer is used really intelligently, this kind of planning can be carried even further.

At the time of year when a seasonal glut makes apples cheap, for instance, buy a really large quantity. One afternoon's work in the kitchen will keep you supplied for many months to come with apple purée for pies and other sweets, apple sauce for meat dishes, fresh sliced apples for cooking as well as using in salads, when this most popular of all fruits is scarce and expensive.

You may notice a special offer of chicken at a lower-than-usual price, either by your usual frozen food supplier, or at a local supermarket. Sometimes one shop brings down the price of chicken to a level that is not really economic, but which tempts many extra customers to do a week's shopping for groceries there, instead of at their usual stores. When chicken is a 'loss leader', as it's called, rush in and take advantage of the bargain. This is the time to go in for chain cooking. Buy a dozen roasters, and a few hours work will provide you with plenty of frozen pieces ready to cook, various prepared chicken dishes that only need thawing and reheating, and lots of good strong stock made from the carcasses.

It may be that you just happen to have some time to spare. No food is specially cheap, but you do have free time and stove space for a big cook-up. A large quantity of minced beef, always a good economy buy for main meal dishes, can be made up into a variety of dishes and sauces. Again, a few hours spent cooking ahead will save you many chores in the months to come.

Minced meat can be shaped into patties and packed in plastic containers with foil or polythene tissue paper dividers. Pat out on a floured board to about ½-inch (1-cm.) thickness and trim to fit the container. Transfer with a fish slice to the container.

MINCED BEEF CHAIN

Make up 7 lb. (3 kg.) of the basic recipe, then by making simple additions you can produce five different dishes – amounting to 40 servings.

Basic meat recipe

2 oz./50 g. dripping or lard

2 lb./900 g. onions, peeled and chopped

3 cloves garlic, crushed

7 lb./3 kg. minced beef

1 2¼-oz./64-g. can tomato purée

2 tablespoons sugar

2 teaspoons salt

½ teaspoon pepper

Heat the fat in a very large saucepan or a preserving pan, then fry the onion and garlic over gentle heat until soft but not brown. Add the meat and fry briskly, stirring constantly to brown. Mix in the remaining ingredients and remove from the heat.

Steak maison

2 lb./900 g. basic meat recipe
2 green peppers, de-seeded and chopped
4 tablespoons tomato ketchup
1 teaspoon salt
2 16-oz./454-g. cans baked beans in tomato sauce

Combine the meat, peppers, ketchup and salt in a saucepan. Cover with a lid and simmer over a low heat for 30 minutes. Add the baked beans.

To freeze: Cool rapidly. Line a meat tin with aluminium foil. Pour the mixture into the tin, cover with foil then partially freeze. Remove from the tin and cut into portions. Quickly re-wrap in double or heavy duty foil and return to the freezer.

To prepare for serving: Unwrap the required number of portions. Reheat in a casserole in the oven or in a tightly covered saucepan. *Makes 2 servings for 4.*

Baboti

1 slice bread, 1½ inches/3½ cm. thick
½ pint/3 dl. milk
2 lb./900 g. basic meat recipe
1 tablespoon Worcestershire sauce
3 tablespoons curry powder

Trim the crusts from the bread, cut into cubes and soak in the milk for 10 minutes. Mix with meat, Worcestershire sauce and curry powder.

To freeze: Pack into foil dishes without pressing the mixture down. Seal with double or heavy duty foil and freeze.

To prepare for serving: Thaw in the refrigerator overnight. Uncover and pour a lightly whisked egg over the meat. Cook in a moderate oven (350°F., 180°C., Gas Mark 4) for approximately 40 minutes. *Makes 2 servings for 4.*

Chilli con carne

1½–2 lb./675–900 g. basic meat recipe
2 teaspoons chilli powder
2 16-oz./454-g. cans baked beans
1 11-oz./312-g. can tomatoes

Combine all the ingredients in a saucepan. Cover with a lid and simmer gently for 40 minutes.

To freeze and prepare for serving: Follow the method of Steak maison.   *Makes 2 servings for 4.*

Shepherd's pie

2 lb./900 g. basic meat recipe
1 tablespoon curry powder
2 oz./50 g. flour
2 tablespoons brown table sauce
¾ pint/4 dl. stock or 1 stock cube and water
3 lb./1¼ kg. mashed potato
salt and pepper to taste

Put the meat into a saucepan and stir in the curry powder, flour and sauce. Cook over gentle heat, stirring constantly, for 5 minutes. Add the stock, bring to the boil and cover with a tight-fitting lid. Simmer gently for 20 minutes.

To freeze: Cool rapidly. Fill aluminium pie dishes or loaf tins two-thirds full. Cover with lightly seasoned mashed potato. Then smooth the tops and seal with double foil and freeze.

To prepare for serving: Remove the foil cover. Reheat in the top of a fairly hot oven (400°F., 200°C., Gas Mark 6) for 30–40 minutes. Remember to fork up the potato after the first 10 minutes of cooking.   *Makes 2 servings for 4.*

Bolognaise sauce

1 lb./450 g. basic meat recipe
1 11-oz./312-g. can tomatoes
¾ pint/4 dl. stock or 1 stock cube and water
¼ pint/1½ dl. red cooking wine (optional)

Combine all the ingredients in a pan. Cover with a tight-fitting lid, simmer gently for about 1 hour.

To freeze: Cool rapidly. Pour into used cream cartons or plastic containers, leaving a ½-inch (1-cm.) headspace. Seal and freeze.

To prepare for serving: Immerse the container in hot water for a few minutes, to loosen the sauce. Turn into a saucepan, cover with a lid and bring to boiling point. *Makes 2 servings for 4.*

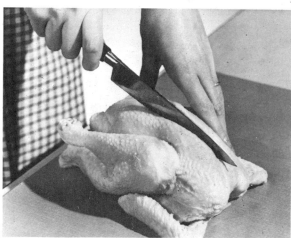

Chicken turnovers, Vol-au-vent filling,
Kromeskies and Chicken pancakes

Freeze eight plain or use for Country house chicken.

For Chicken curry use:

★ Two chickens jointed into 12 portions.

For Chickenburgers use:

★ The meat from one boned chicken.

For Chicken turnovers, Kromeskies, Vol-au-vent, and Pancakes use:

★ Three chickens, roasted.

For Chicken stock use:

★ All the carcasses.

Chicken pieces, plain and coated

CHICKEN CHAIN

The recipes in this section show just how versatile chicken can be. Buy a dozen roasters at a time and use them in the following way:

Freeze two whole.

For Chicken Kiev use:

★ The breasts and wings of three chickens.

For Chicken pieces use:

★ The remainder of the three chickens, plus an additional one, cut into joints, making 12 pieces in all. Egg and breadcrumb four, wrap in double foil or polythene tissue paper, or place in foil dishes and seal with double thicknesses of foil.

Thaw the bird sufficiently to remove the giblets from the body cavity. Place on a chopping board and with a sharp knife cut through and along the length of the breastbone.

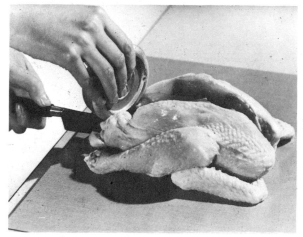

Open the bird out, then cut through along the length of the backbone. If liked, the backbone can be removed entirely by cutting along close to either side. Tap the back of the knife sharply with a heavy weight to cut the bird in half through the bony sections.

100

Lay the chicken halves skin side up on a board and halve again by cutting diagonally across between wing and thigh. The bird is now in quarters.

To make six joints divide each thigh and drumstick portion in half by cutting through at the ball and socket joint.

Chicken curry

12 chicken pieces
seasoned flour, for coating
2 oz./50 g. lard or dripping
2 onions, peeled and chopped
1 pint/6 dl. curry sauce (see page 83)

Coat the chicken pieces in the seasoned flour. Melt the fat in a large, heavy frying pan; add the onion and cook until transparent. Transfer to a baking dish. Add the chicken pieces, a few at a time, to the fat in the pan and cook until brown on all sides.

Place the chicken in the dish with the onion, add the sauce, cover with foil. Bake in a moderate oven (350°F., 180°C., Gas Mark 4) for 1 hour.

To freeze: Cool rapidly and pack into suitable containers.

To prepare for serving: Reheat, unthawed, in a saucepan. Serve with plain boiled rice. *Makes 12 portions.*

Chickenburgers

3 slices white bread
1 lb./450 g. shoulder of veal, skinned
1 3-lb./1¼-kg. chicken, boned
¼ pint/1½ dl. milk
½ teaspoon salt
¼ teaspoon nutmeg
1 large egg

Soak the bread in the milk. Trim the veal and cut the chicken into pieces. Mince the meats finely. Mix in the bread and all the remaining ingredients. Divide into 12 portions and shape into patties.

To freeze: Wrap in double or heavy duty foil. Use dividers between the layers if to be stacked. Seal and freeze.

To prepare for serving: Shallow-fry thawed or frozen, in plenty of butter or oil. *Makes 12.*

Chicken turnovers

12 oz./350 g. puff pastry
12 oz./350 g. cooked chicken, finely chopped
1 small onion, finely chopped
1 10½-oz./298-g. can condensed mushroom soup
1 tablespoon chopped parsley, blanched
½ teaspoon salt
dash Worcestershire sauce

Roll the pastry to ⅛-inch (3-mm.) thickness. Cut into 5-inch (13-cm.) circles, using a saucer. Combine the remaining ingredients and put a spoonful of filling into the centre of each pastry circle. Damp the edges and seal with the prongs of a fork.

To freeze: Pack into foil or plastic containers. Cover closely and seal before freezing.

To prepare for serving: Put the turnovers, unthawed, on a wet baking sheet, brush with beaten egg and cook in a very hot oven (450°F., 230°C., Gas Mark 8) for 30 minutes. *Makes 12.*

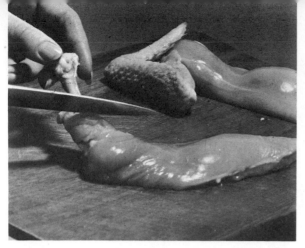

Scrape back the meat from the wing bone.

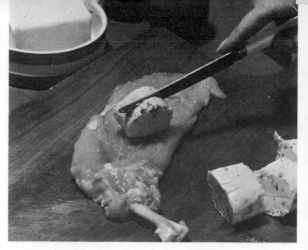

Place a portion of butter in the centre of each breast.

Dip in beaten egg and coat with crumbs.

Chicken Kiev

Chicken Kiev

3 whole chicken breasts, with wings attached
6 oz./175 g. parsley butter
salt and pepper to taste
seasoned flour
1 egg, lightly beaten
toasted breadcrumbs for coating

To remove the chicken breasts, skin the chicken and place with the front facing you. Using a sharp filleting knife, carefully remove the flesh from one side of the breast bone, without piercing the meat. Gradually work the tip of the knife down towards the wing. Sever the tendon between the ball and socket wing-joint and cut through the skin underneath it. Repeat this process on the other side of the chicken. Treat the two other chickens in the same way. Cut off the two end bones from each wing, scrape back the meat from the remaining bone (see pictures).

Put each chicken breast between two sheets of wet greaseproof paper and pound until thin, taking care not to split the meat. Roll the butter, cut into six equal portions and place one in the centre of each breast.

Sprinkle with salt and pepper. Roll up envelope fashion, letting the wing bone protrude. Toss each roll in well-seasoned flour; dip into the egg and coat with crumbs.

To freeze: Pack in a plastic container, cover closely and freeze.

To prepare for serving: Thaw overnight in the refrigerator. Deep-fry in hot oil, drain and serve. Garnish with parsley and top with a little extra parsley butter if liked. *Serves 6.*

Kromeskies

1 lb./450 g. cooked chicken
8 oz./225 g. cooked ham or bacon
3 oz./75 g. margarine
1 large onion, peeled and finely chopped
1 clove garlic, crushed
3 oz./75 g. flour
¾ pint/4 dl. chicken stock or 1 stock cube and water
2 tablespoons tomato ketchup
salt and pepper to taste
¼ teaspoon powdered mace
12 rashers streaky bacon, trimmed

Put the chicken and ham through a fine mincer. Melt the margarine in a saucepan and add the onion and garlic. Cook gently until soft but not brown. Stir in the flour and cook for 1 minute, gradually add the stock. Add all the other ingredients, with the exception of the bacon. Bring to the boil and cook gently, beating with a wooden spoon, until the mixture begins to leave the sides of the pan. Spread into a sandwich tin and leave to cool. Meanwhile, stretch the bacon by smoothing it with a dinner knife and cut each rasher in half. When the mixture is quite cold, divide it into 24 equal portions. Flour both hands and roll each portion into a ball and then into a cork shape. Wrap a piece of bacon around each.

To freeze: Arrange in oblong plastic or foil containers. Stack layers on top of each other, with foil or polythene tissue paper dividers between. Seal and freeze.

To prepare for serving: Thaw at room temperature, allowing approximately 2 hours. Toss in seasoned flour, dip in coating batter and deep-fry in hot fat (380°F., 193°C.) for 2–3 minutes. *Makes 3 servings for 4.*

Chicken stock

To each 1 lb./450 g. chicken bones and trimmings use:
3 pints/1¾ litres water
1 small carrot, sliced
1 small onion, stuck with a clove
small bouquet garni

Wash the carcasses and trimmings. Place in a large saucepan and add the water. Bring slowly to the boil and simmer for 2 hours. Skim frequently to remove the fat from the surface of the stock and the sides of the pan. Add the remaining ingredients and enough cold water to replace that which has evaporated. Simmer gently for another hour. Skim again and strain the stock through a muslin cloth.

Put the stock into a clean pan; add the vegetables and bouquet garni. Return to the boil and simmer for another 30 minutes. Remove any remaining scum with kitchen paper.

To freeze: Cool and strain into plastic containers or polythene bags leaving a 1-inch (2½-cm.) headspace. Separate cubes, frozen in an ice cube tray from the refrigerator, supply just enough strong stock to improve a gravy or sauce.

To prepare for serving: Heat in a saucepan and use as a base for soups or sauces.

Vol-au-vent filling

1½ pints/scant litre milk
1 small onion
blade mace
4 oz./100 g. butter or margarine
3 oz./75 g. flour
1 green pepper, de-seeded
8 oz./225 g. button mushrooms, quartered
1¼ lb./575 g. cooked chicken, diced
salt and pepper to taste

Warm the milk with the onion and mace. Allow to infuse for 30 minutes, then strain. Melt 3 oz. (75 g.) of the butter in a saucepan; stir in the flour and cook gently for a further 2 minutes. Draw the pan off the heat and whisk in the milk. Blanch the green pepper in boiling water for 7 minutes. Immerse in cold water to cool. Cut into ¼-inch (½-cm.) dice.

Heat the remaining butter in a small saucepan; add the pepper and mushrooms and cook for 1½ minutes. Add, together with the chicken, to the sauce. Return the pan to the heat, bring to the boil and cook for a further 2 minutes, stirring continuously. Season to taste.

To freeze: Cool rapidly and pack into a large plastic bowl if required for a forthcoming party, or into several small containers if the filling is required for family meals.

To prepare for serving: Thaw either in the refrigerator or at room temperature. The time required depends on the size of the container. Reheat in a saucepan before filling the vol-au-vent cases. *Fills 12 large cases.*

Country house chicken

8 chicken pieces

2 tablespoons seasoned flour

2 oz./50 g. butter

8 oz./225 g. button mushrooms

2 chicken stock cubes

¾ pint/4 dl. boiling water

4 tablespoons dry sherry

2 teaspoons cornflour

1 tablespoon orange jelly marmalade

Turn the chicken pieces in the seasoned flour. Heat half the butter in a flameproof casserole and sauté the chicken pieces gently until golden brown all over. Cover and continue cooking gently, turning frequently, for 20 minutes. In a separate pan, sauté the mushrooms in the remaining butter for 3 minutes, slicing the larger ones. Dissolve the stock cubes in the boiling water, stir in sherry, cornflour moistened with a little cold water and the marmalade. Add the mushrooms to the chicken, strain the liquid over them. Bring to the boil, cover and simmer for a further 5 minutes. Cool.

To freeze: Pack in plastic containers. Seal and label.

To prepare for serving: De-frost, turn into a saucepan, reheat gently to boiling point and simmer for 10 minutes. Stir frequently to prevent burning. *Serves 4–6.*

Note: This dish is also excellent made with frozen chicken for re-freezing or for serving immediately.

Basic pancakes

8 oz./225 g. plain flour

½ teaspoon salt

4 eggs

1 pint/6 dl. milk

4 teaspoons corn oil

oil for frying

Sieve the flour and salt into a basin. Lightly beat the eggs, add to the dry ingredients and gradually beat in the milk, until the batter is smooth. Finally beat in the corn oil. Heat a little oil in a small pan and use to fry small thin pancakes until golden brown on both sides.

To freeze: Cool and pack together in layers with polythene tissue, double or heavy duty foil dividers. Place in a polythene bag. Seal and label.

To prepare for serving: Uncover and thaw at room temperature for 30 minutes, spread out. To reheat, place on baking trays and put into a moderately hot oven (400°F., 200°C., Gas Mark 6) for 10 minutes. *Makes about 36.*

Chicken pancakes

For the filling:

2 oz./50 g. butter

1 large onion, peeled and finely chopped

2 oz./50 g. flour

1 pint/6 dl. chicken stock or milk

1 lb./450 g. cooked chicken, finely minced

salt and pepper to taste

dash Worcestershire sauce

36 small thin pancakes

1 pint/6 dl. tomato sauce (see page 83)

12 oz./350 g. cheese, grated

Melt the butter in a saucepan, add the onion and cook until soft. Gradually stir in the flour and cook gently for 2 minutes, add the stock or milk, bring to the boil and cook for a further 2 minutes, stirring constantly. Stir in the remaining ingredients. Allow the mixture to cool, then spoon a little of the filling into each thawed pancake. Roll and pack closely in foil-lined dishes. Cover with tomato sauce (see page 83) and sprinkle with grated cheese.

To freeze: Cover with foil and partially freeze. When the pancakes are sufficiently solid, lift out of the dishes and wrap in double or heavy duty foil. Seal and return to the freezer.

To prepare for serving: Unwrap the pancakes and return to their original dishes. Thaw at room temperature allowing approximately 5 hours. Cook in a moderately hot oven (375°F., 190°C., Gas Mark 5) for 30–40 minutes. *Makes 3–4 servings for 4.*

APPLE CHAIN

Buy 16 lb. (7 kg.) cooking apples. If they are slightly damaged, buy a few extra pounds and discard the bruised parts. Use half for the Basic apple purée recipe, then pack in ¼- or ½-pint (1½- or 3-dl.) containers. These small packs can be thawed quickly without waste for use in puddings, sauces or as baby food.

Use the remainder of the apples sliced. A few can be stored with blackberries – by the dry pack method – for pies or to serve as stewed fruit. Store some slices in plain syrup packs and some unsweetened, but for convenience store the largest portion in dry sugar packs. Use these for puddings and pies.

Basic apple purée

8 lb./3½ kg. cooking apples
½ pint/3 dl. water
2 oz./50 g. butter or margarine
juice and zest of 1 lemon
8 oz./225 g. granulated sugar

Wash and quarter the apples. Put them into a large saucepan or preserving pan without peeling or coring. Add the water, butter or margarine, lemon juice and zest. Cover with a lid or piece of aluminium foil. Cook gently, shaking the pan from time to time. When the apples are almost pulped add the sugar and then beat with a wooden spoon until completely smooth. Continue to cook until the sugar is completely dissolved. Rub the mixture through a fine hair or nylon sieve and discard the peel and cores.

To freeze: Pour the purée into cream cartons, plastic or other suitable containers, leaving a ½-inch (1-cm.) headspace. For convenience, pack in small quantities, just enough for one serving or to suit your family's needs. Seal and freeze.

To prepare for serving: Either thaw at room temperature (the length of time required depends on the size and shape of the container) or place the container under hot running water to loosen. Turn the contents into a saucepan and heat.

Packing basic apple purée

Apple snow

sugar to taste
½ pint/3 dl. basic apple purée, thawed
green colouring
2 egg whites

Add the sugar to the apple purée and tint with a little green colouring but be careful not to add too much. Whisk the egg whites stiffly and fold into the apples. Spoon into individual dishes and chill; and, if liked, garnish with apple slices before serving. *Serves 4.*

Apple snow

Apple fool

¼ pint/1½ dl. custard, cold
½ pint/3 dl. basic apple purée, thawed
sugar to taste

Mix the custard into the apple purée. Sweeten to taste and spoon into individual dishes. Chill before serving.   *Serves 4.*

Baked apple soufflé

3 oz./75 g. castor sugar
pinch ground cinnamon
½ pint/3 dl. basic apple purée, thawed
2 large eggs, separated

Mix the sugar and cinnamon into the apple purée then beat in the egg yolks. Whisk the egg whites stiffly and fold into the mixture using a metal spoon. Spoon into a heavily greased 1½-pint (1-litre) soufflé or pie dish. Bake in the centre of a moderate oven (350°F., 180°C., Gas Mark 4) for 30–40 minutes or until firm to the touch and golden brown. Serve at once with cream or custard.   *Serves 4.*

Apple cheese flan

6 oz./175 g. shortcrust pastry
2 eggs, lightly whisked
3 oz./75 g. castor sugar
finely grated rind of ½ lemon
¼ pint/1½ dl. basic apple purée, thawed

Line an 8-inch (20-cm.) flan or loose-based sandwich tin with the pastry. Prick the base and bake blind, in a pre-heated hot oven (400°F., 200°C., Gas Mark 6) for 15 minutes. Meanwhile, beat the eggs, sugar and lemon rind into the apple purée.

Remove the flan from the oven and lower the heat to 350°F., 180°C., Gas Mark 4. Pour the filling into the case. Continue to cook for 30–35 minutes or until the filling is firm. Serve hot or cold.   *Serves 6.*

Apple slices

Peel and core 8 lb. (3½ kg.) apples and cut into ½-inch (1-cm.) slices. Blanch in boiling water for 1 minute, and pack into polythene bags or other suitable containers. Seal and freeze.

To prepare for serving: Partially thaw in the sealed container, at room temperature.

Packing apple slices

Apple and sultana layer pudding

3 oz./75 g. self-raising flour
pinch salt
¼ teaspoon mixed spice
1½ oz./40 g. white breadcrumbs
2 oz./50 g. shredded suet
3–4 tablespoons milk
For the filling:
8 oz./225 g. apple slices, partially thawed
2 oz./50 g. sultanas
2 oz./50 g. brown sugar
rind of ½ lemon, finely grated

Sieve the flour, salt and mixed spice into a mixing bowl. Stir in the breadcrumbs and suet. Add sufficient milk to make a soft, scone-like dough. Divide into three unequal portions. Prepare the filling by finely chopping the apple slices. Mix with the remaining ingredients. Grease a 1-pint (6-dl.) pudding basin and pat out the smallest piece of suet pastry to fit the base of the bowl. Add half the fruit filling and pat out the second piece of pastry to cover this. Add the rest of the filling and pat out the remaining pastry to cover it. Cover the basin with a piece of foil. Steam for 1½–1¾ hours. Serve with custard.   *Serves 4.*

Apple charlotte

Apple pie

8 oz./225 g. shortcrust pastry

12 oz./350 g. apple slices, partially thawed

2 oz./50 g. castor sugar

3 whole cloves

To decorate:

beaten egg white or milk

castor sugar

Divide the pastry into two pieces, one slightly smaller than the other. Knead each to a smooth ball. Roll the smaller piece very thinly, to fit an 8–8½-inch (20–21-cm.) pie plate. Cover the plate with the pastry and press out all the air bubbles, starting from the centre.

Put a thick layer of apple over the pastry; sprinkle with castor sugar and add the cloves. Pile the remaining apple slices on top and smooth to a flat dome. Damp the edges of the pastry. Roll the remaining pastry and, without stretching it, cover the top of the pie. Press the edges together to seal; trim and flute. Make a small vent in the centre, brush with milk or lightly beaten egg white and sprinkle with castor sugar. Bake in the centre of a moderate oven (350°F., 180°C., Gas Mark 4) for 30–40 minutes. *Serves 4.*

Apple charlotte

12 oz./350 g. apple slices, partially thawed

3 oz./75 g. breadcrumbs

1 oz./25 g. brown sugar

1 tablespoon golden syrup

grated rind and juice of ½ lemon

Grease a ¾-pint (4-dl.) pie dish and arrange the apple slices and breadcrumbs in alternate layers, starting and finishing with breadcrumbs. Warm the remaining ingredients in a small saucepan, to blend. Spoon the syrup over the crumbs. Bake in the centre of a warm oven (325°F., 170°C., Gas Mark 3) for 50–60 minutes. Garnish with grated lemon rind. *Serves 4.*

Apple blintzes

8 fl. oz./2¼ dl. milk
4 eggs
1 teaspoon salt
4 oz./100 g. plain flour
oil, for greasing pan
For the filling:
1½ lb./675 g. cottage cheese, drained
pinch salt
½ oz./15 g. butter
2 tablespoons castor sugar
8 oz./225 g. apple slices, partially thawed

To make the batter, beat together lightly the milk, salt and eggs. Beat in the flour until smooth. Heat a griddle or heavy shallow pan and grease lightly with oil. Spread about 2 tablespoons of the batter evenly over the griddle and cook until bubbles form on the surface. Turn and cook the other side.

To make the filling, press the cheese through a sieve, or mash with a fork. Add salt to taste, add the butter and castor sugar and beat well together. Place one or two apple slices and a tablespoon of filling in the centre of each pancake and fold envelope style.

To freeze: Pack in a plastic container in layers with polythene tissue, double or heavy duty foil dividers. Seal and freeze.

To prepare for serving: Thaw. Fry in melted butter until golden brown on both sides. Serve with soured cream into which an equal quantity of warmed apricot jam has been stirred. *Serves 4–6.*

Eve's pudding

12 oz./350 g. apple slices, partially thawed
3 oz./75 g. castor sugar
4 oz./100 g. plain flour
1½ teaspoons baking powder
pinch salt
2 oz./50 g. margarine
1 egg
¼ teaspoon vanilla essence
2–3 tablespoons milk

Mix the apple slices with 1 oz. (25 g.) of the castor sugar and arrange them in a 1-pint (6-dl.) pie dish. Sieve the flour, baking powder and salt. Add the remaining sugar and fat, cut into pieces, to the flour. Rub in until the mixture resembles fine breadcrumbs. Beat the egg lightly with the vanilla essence. Mix into the fat and flour with sufficient milk to make a creamy consistency.

Spread the mixture over the surface of the apples and bake in a moderate oven (350°F., 180°C., Gas Mark 4) for 40 minutes. *Serves 4.*

Chapter Nine

Buffet parties

This kind of party has long been popular as the relaxed, easy way to entertain. The guests do most of the work once the party has started, helping themselves to plates, cutlery and food. However, it does presuppose that the hostess has lots of time on her hands for preparing the food beforehand.

You, the fortunate freezer owner, can spend an hour or two on several different days making up the provisions for the buffet table! You can also serve a frozen sweet straight from the freezer to add a dramatic touch to the festivities when the more solid titbits have disappeared and appetites are flagging. How else could you produce 15 or 20 frozen Orange cups, each nestling in the skin of the orange and gleaming with frost, whenever you want to? Most of the recipes in the following section serve 15, as this is a fair average number for such a party.

If you are serving 15 guests, a selection of seven recipes should be adequate in quantity – two sweet, four cold savoury and one hot savoury should be an ideal balance of dishes. For fewer numbers, drop one savoury recipe, and for more, increase the selection, bearing in mind that most guests will average seven selections from the buffet table. (Greedy guests will eat more, but then there may be a few who will be too busy talking to eat as many.)

Another advantage which the freezer bestows is that you can prepare more than you think you will need and only completely thaw sufficient to set a generous table at the start of the party. If the food vanishes more rapidly than you had expected, you produce your extra supplies from the refrigerator. But if they are not needed, they can remain happily in the freezer for another day.

Onion bread rolls

½ pint/3 dl. tepid water
½ oz./15 g. dried yeast
1 teaspoon sugar
2 tablespoons corn oil
1 lb./450 g. plain flour
1 packet French onion soup

Pour the water on to the yeast. Add the sugar and leave until the yeast has dissolved. Stir in the corn oil and mix well. Sieve the flour into a mixing bowl and add the contents of the packet of soup. Stir in the liquid and mix to form a soft dough, adding a little more water if necessary. Beat the dough until it leaves the sides of the bowl clean. Cover with a damp cloth and leave to rise in a warm place, until double in bulk. Turn out and knead lightly.

Divide the dough into 15 equal portions and shape into rolls. Place on a greased baking sheet and leave to prove in a warm place for 10–15 minutes. Bake in a hot oven (425°F., 220°C., Gas Mark 7) for 10–15 minutes.

To freeze: Cool and pack in polythene bags. Seal tightly and freeze.

To prepare for serving: Thaw at room temperature in the sealed bag. *Makes 15.*

Hamburger patties

2 lb./900 g. lean chuck steak
1 onion, peeled
1 clove garlic
2 slices bread
1 egg
1 teaspoon salt
pinch pepper
¼ teaspoon sage

Remove all fat from the steak, mince it twice with

the onion and garlic, using a coarse cutter. At the end of the second mincing, add the bread, so that this goes through the mincer last. Turn the meat into a basin, add the beaten egg and seasoning. Mix well. Form into cakes about 3 inches (7½ cm.) across and ¼ inch (½ cm.) thick.

To freeze: Arrange four patties in a row, put a dividing piece of foil on top; continue to stack with foil dividers between each layer. Wrap tightly in double foil, seal and freeze.

To prepare for serving: Partially thaw in the refrigerator, allowing 5 hours. Pat dry mustard into the surface before grilling. Baste well with barbecue sauce while cooking. Sandwich hot patties between halves of soft rolls with more mustard and any good pickles, or relishes, or thin slices of onion. *Makes 16–18 patties.*

Savoury sandwich loaf

1 large white loaf, unsliced
2 eggs, hard-boiled
¼ pint/1½ dl. single cream
2 oz./50 g. butter
salt and pepper
squeeze lemon juice
green colouring
1 tablespoon spinach purée
1 2-oz./57-g. jar lobster paste
3 packets demi-sel cheese

Remove the outside crust from the bread and then slice lengthways horizontally into four. Using a fork, mash the hard-boiled eggs, while still warm, with 1 tablespoon of the cream and ½ oz. (15 g.) of the butter. Season well. Beat the remaining butter with a squeeze of lemon juice to soften and add a little green colouring. Add the spinach purée. Spread the first piece of bread with the lobster paste, and cover with a second piece. Spread with the egg mixture and cover with the third layer. Spread this one with the spinach butter and cover with the top layer of bread. Blend together the demi-sel cheese and the remaining cream until soft and of a spreading consistency. Completely cover the loaf with the cream cheese mixture and smooth with a palette knife. Chill and slice, using a sharp knife. Reshape into a loaf.

110

To freeze: Pack in a suitable plastic container. Seal and freeze.

To prepare for serving: Thaw, covered, at room temperature, allowing approximately 5 hours. Cut each slice in half lengthways. *Makes 15 slices.*

Sausage rolls

2 7-oz./198-g. packets puff pastry
12 oz./350 g. pork or beef sausage meat
salt and pepper

Roll one packet of pastry thinly to approximately 12 by 6 inches (30 by 15 cm.). Trim the edges with a knife and cut in half lengthways. Season the sausage meat well, divide in half and roll one portion into two long rolls the same length as the pastry. Place down the centre of each strip. Brush one side of the pastry with a little water, fold over and seal. Using a sharp knife, cut each strip into six sausage rolls. Make up the other packet of pastry and the remaining sausage meat in the same way.

To freeze: Chill until the pastry is firm. Stack in layers, with foil or polythene tissue dividing papers. Seal and freeze.

To prepare for serving: Place the frozen sausage rolls on wet baking sheets. Cook in a very hot oven (450°F., 230°C., Gas Mark 8) for 10 minutes. Reduce the temperature to 425°F., 220°C., Gas Mark 7 for a further 10 minutes. *Makes 24.*

Cheese and apple horns

1½ lb./675 g. shortcrust pastry, flavoured with
    1 oz./25 g. Parmesan cheese, grated
1 egg, beaten
1 oz./25 g. extra Parmesan cheese, grated

Roll the pastry to an oblong and trim to 14 by 4½ inches (35 by 10 cm.), then cut into six strips each ¾ inch (2 cm.) wide. Wind overlapping around the horn tins. Place on a baking sheet, with the pastry ends underneath. Brush with egg, sprinkle with the extra Parmesan cheese. Bake in a fairly hot oven (400°F., 200°C., Gas Mark 6) for 15–20 minutes. Cool on a wire tray.

To freeze: Pack in a large plastic container. Seal and freeze.

To prepare for serving: Thaw, covered, at room temperature allowing approximately 5 hours. Fill with grated apple and cream cheese. *Makes 18–20.*

Cheese and apple horns

Tuna pâté

Tuna pâté

2 7-oz./198-g. cans tuna steak

little milk

blade mace

small piece bay leaf

sprig parsley

4 black peppercorns

1 oz./25 g. butter

2 oz./50 g. flour

4 tablespoons single cream or top of milk

1–2 teaspoons lemon juice

few gherkins, finely chopped (optional)

Drain the liquor from the tuna and make up to ½ pint (3 dl.) with milk. Place in a saucepan with the mace, bay leaf, parsley sprig and peppercorns. Leave over a gentle heat to infuse for 15 minutes,

then bring to the boil and strain. Melt the butter in a small saucepan and stir in the flour. Cook for 2–3 minutes, add the milk mixture and bring to the boil, stirring; cook for 2 minutes. Mash the tuna fish with a fork, add to the sauce with the cream, lemon juice and gherkins and mix well together. Add additional seasoning if necessary. Place in a foil-lined serving dish and leave in a cool place to set.

To freeze: Cover with foil and partially freeze. When sufficiently solid, remove from the container and over-wrap with double foil.

To prepare for serving: Unwrap and return to the original serving dish. Thaw at room temperature allowing approximately 2 hours. Decorate with parsley and lemon. *Serves 16.*

Kidney risotto

3 tablespoons cooking oil

2 large onions, peeled and thinly sliced

2 cloves garlic, crushed with 2 teaspoons salt

2 lb./900 g. Patna rice

3 pints/1¾ litres chicken stock or 3 stock cubes and water

3 green peppers, de-seeded and sliced

2 oz./50 g. butter

8 lambs' kidneys, trimmed and sliced

12 oz./350 g. mushrooms, quartered

Heat the oil in a large saucepan, add the onions and garlic and cook over a low heat until soft but not brown. Add the rice and fry for 2 minutes before adding the stock. Bring to the boil, cover with a lid and simmer, very gently, for 10 minutes. Add the peppers and cook for a further 10 minutes. Heat the butter in a frying pan and gently sauté the kidneys; remove from the pan and replace with the mushrooms, cook for 2 minutes.

To freeze: Cool all the ingredients, then toss the kidneys and mushrooms in the rice. Spoon into large plastic containers or polythene bags, seal and freeze.

To prepare for serving: Immerse containers in hot water to loosen the contents. Turn into meat dishes, cover with foil and reheat in a warm oven (325°F., 170°C., Gas Mark 3). Place in serving dishes, sprinkle with grated cheese. *Serves 15.*

Arrange alternate layers of lasagne, cheese and sauce and sprinkle the top with grated cheese.

The trays can either be frozen separately or with one turned upside down on top of another. If you up-turn them, only one need be covered with foil. Partially freeze, then place together.

Seal the joining edges of the two trays with freezer tape. Don't use ordinary adhesive tape as it doesn't stand up to low temperatures.

As you go round the edge with the tape, pinch the trays together to make sure the tape sticks properly to ensure a good seal.

Lasagne

1 2 lb. 3-oz./992-g. can tomatoes
1 pint tomato sauce (see page 83)
1 teaspoon salt
¼ teaspoon pepper
4 tablespoons oil
2 large onions, sliced
1 clove garlic, crushed
2 lb./900 g. minced beef
2 1-lb./454-g. packets lasagne
3 8-oz./226-g. cartons cottage cheese
4 oz./100 g. Parmesan cheese, grated

Combine the tomatoes, tomato sauce, salt and pepper in a saucepan. Bring to simmering point and cover with a lid. Heat 3 tablespoons of the oil in a frying pan, add the onion and garlic and cook until soft but not brown. Stir the meat into the onion and cook briskly to brown. Spoon the meat into the tomato sauce and simmer, uncovered, for 1 hour. Add the remaining oil to a large saucepan (or preserving pan) of salted, boiling water. Add the lasagne and cook, at boiling point, for 12 minutes. Drain and separate the pieces.

To freeze: Use foil dishes or line serving dishes with foil, spoon a little meat sauce into the base of each. Divide the cottage cheese, allowing equal proportions for each dish. Arrange alternate layers of lasagne, cottage cheese and sauce, finishing with the sauce. Sprinkle the top of each with grated cheese. Cover with foil and partially freeze. Remove from the containers and re-wrap in double or heavy duty foil. Return to the freezer.

To prepare for serving: Remove the foil and transfer to the original serving dishes. Thaw at room temperature. Cook in a moderate oven (350°F., 180°C., Gas Mark 4) for 30–40 minutes. *Serves 15.*

Party tartlets

Bake blind your usual rich shortcrust pastry tartlet cases. When cold, stack them together, place them in suitable containers, seal, label and freeze. When you are ready to use them, remove from the freezer and allow them to come to room temperature before crisping a little in the oven. Now add a hot savoury or sweet filling.

Raw pastry tartlet cases can be kept equally well in the freezer and are simply baked when required. However, for a party you will save a little time by baking them before freezing.

Small fruit tartlets are also excellent for buffet parties. Instead of a rich shortcrust pastry, use a biscuit crust mixture to make them more unusual. Here is a good basic savoury filling.

Basic savoury filling

1 oz./25 g. butter

1 tablespoon flour

¼ pint/1½ dl. chicken stock

2–3 oz./50–75 g. button mushrooms, washed but unpeeled

small nut butter

1 tablespoon water

1 teaspoon lemon juice

6 oz./175 g. cooked chicken or turkey, cut finely

2 tablespoons cooked peas

1 4½-oz./128-g. can pimento, cut in thin strips

1 inch/2½ cm. long and drained

Melt the butter and stir in the flour to make a smooth roux. Remove from the heat, stir in chicken stock and cook for a few minutes. Meanwhile, slice the mushrooms and cook them, covered, in a little butter, water and lemon juice. Add the mushrooms to the sauce together with the chicken or turkey. Add the peas and pimento. Heat through. Fill the tartlet cases straight away and serve while still hot.

This filling can be made in advance and freezes well. When you are ready to use it, allow to thaw and then heat through before filling the warmed tartlet cases.

The basic savoury mixture can be varied in several ways. For instance, instead of chicken or turkey, use shelled shrimps or a can of crab meat, flaked, or tuna, or salmon. Or, if you have any cooked salmon in the freezer, flake it roughly and use as the basis of the recipe.

Orange cups

3 lemons

10 oranges, halved

water

1 lb./450 g. granulated sugar

4 egg whites

Pare the skin off one lemon and five orange halves. Squeeze out and reserve the juice of all the oranges and lemons, taking care not to split the orange cases. Set 15 orange halves aside. Add water to the fruit juice to make 2½ pints (1½ litres) and place in a large saucepan with the pared skin. Dissolve the sugar in the liquid over gentle heat. Bring to the boil and continue to boil rapidly, without stirring, for 12 minutes. Cool and strain into a large mixing bowl. Freeze for 6 hours. To prepare the orange cases, insert the handle of a teaspoon between the pith and skin of each orange. Loosen the pith all the way round each case. Discard the pith. Break the prepared sorbet into pieces with a fork, and then mash. Whisk the egg whites stiffly and fold into the sorbet. Spoon into the prepared cases and return to the freezer with the remaining sorbet. Top each orange case with a scoop of the remaining sorbet before serving. *Serves 15.*

Pear Hélène tartlets

3 oz./75 g. castor sugar

½ pint/3 dl. water

4 pears, peeled, cored and cut into quarters

juice of ½ lemon

8 oz./225 g. plain chocolate, melted

1 tablespoon oil

ice cream, for filling

15 frozen tartlet cases, thawed

Dissolve the sugar in the water, bring to the boil and continue to boil rapidly for 5 minutes. Lower the heat and simmer the pears in the syrup until tender, do not allow them to break. Drain, cool and sprinkle with lemon juice.

Add the oil to the melted chocolate and keep it warm.

Just before serving, put a scoop of ice cream in each case, top with a piece of pear and drizzle melted chocolate over. *Makes 15.*

Gingered pear coupes

1 quantity buttered crumbs (see below)
3 16-oz./454-g. containers frozen sliced pears in sugar
   syrup
6 tablespoons ginger syrup
¾ pint/4 dl. double cream
sugar to taste
3 egg whites
12 small pieces stem ginger

De-frost the containers of buttered crumbs and pears in warm water. Divide half the crumbs between 12 coupe glasses. Arrange the pear slices upright over the crumb base, and pour over the ginger syrup and a little syrup from the frozen pears. Fill up the coupes to the level of the top of the pear slices with more buttered crumbs. Whip the cream and sweeten to taste. Fold in the stiffly beaten egg white and spoon or pipe cream over the crumbs and fruit. Decorate with the stem ginger. *Serves 12*.

Sweet buttered crumbs

These can be used in a number of recipes with fruit and fruit purées. Make up a basic quantity as follows and freeze in four small plastic containers.

4 oz./100 g. unsalted butter
12 oz./350 g. dried white crumbs
4 oz./100 g. demerara sugar

Melt the butter, stir in the breadcrumbs and fry until beginning to turn brown. Add the sugar and continue frying until the mixture is golden brown, stirring constantly. Cool.

To freeze: Pack in plastic containers.

To prepare for use: Thaw in warm water.

Pear condé

2 15-oz./425-g. cans creamed rice pudding
2 30-oz./850-g. containers pear halves in sugar syrup
4 tablespoons apricot jam
½ pint/3 dl. double cream
16 chocolate leaves (see note)

Place the creamed rice in the base of two shallow serving dishes. De-frost the containers of pears in warm water. Drain the pears well and arrange, cut side down, over the rice. Heat the pear syrup with the jam, allow to cool slightly and strain over the pears to glaze. Whip the cream, pipe a border of rosettes all around the dish and between the pear halves. Decorate with chocolate leaves. *Serves 15*.

Note: To make chocolate leaves, melt plain chocolate in a small bowl over hot water. Choose even-size rose leaves and wash and dry carefully. Using the back of a teaspoon coat the underside of the leaves with melted chocolate. Allow to cool until quite hard then gently ease off the leaves.

Chocolate and lime crunch pie

1 oz./25 g. butter
4 tablespoons golden syrup
8 oz./225 g. cornflakes
6 oz./175 g. plain chocolate, melted
For the filling:
2 packets lime jelly
¾ pint/4 dl. evaporated milk
few drops green food colouring
To decorate:
whipped cream
angelica leaves

Melt the butter and golden syrup together. Add to the cornflakes with the melted chocolate. Toss together lightly until well coated. Line two 8-inch (20-cm.) pie plates and allow to set. Meanwhile dissolve the jellies in ½ pint (3 dl.) boiling water and allow to cool. When nearly at setting point, whisk the evaporated milk until thick and frothy. Fold in the jelly and a little green colouring. Pour into the flan cases and allow to set. Decorate with rosettes of whipped cream and a few angelica leaves.

To freeze: Open freeze until firm, wrap in foil and label.

To prepare for serving: Allow to de-frost, unwrapped, at room temperature. *Serves 12–15*.

Pineapple and mallow krispie pie

2 oz./50 g. butter
4 tablespoons golden syrup
4 oz./100 g. rice krispies

For the filling:

12 oz./350 g. marshmallows
8 tablespoons pineapple juice
few drops yellow food colouring
2 small cans pineapple cubes, drained
½ pint/3 dl. double cream

To decorate:

whipped cream
few pineapple pieces

Melt the butter and the golden syrup in a pan. Add to the rice krispies and mix well. Press the mixture into two 7-inch (18-cm.) pie plates and allow to cool and harden. Melt the marshmallows in a bowl over hot water with the pineapple juice. Cool and add a little yellow colouring. Whisk the cream until thick and fold into the mallow mixture with the pineapple cubes, reserving a few for decoration. Pour into the flan case and allow to set. Decorate with a little whipped cream and a few pineapple cubes.

To freeze: Open freeze until firm, wrap in foil and label.

To prepare for serving: Allow to de-frost, unwrapped, at room temperature. *Serves 12–15.*

Chapter Ten

Teatime entertaining

Unexpected guests have an uncanny knack of dropping in for tea on the one day when you have nothing in the biscuit tin, and no freshly baked cakes to offer them. Such problems can be solved triumphantly with some short-term items from the freezer.

By this I mean such things as scones and tea breads, which you would not expect to keep frozen long, because they are rather bulky and have a relatively short freezer life. Cooked scones, for instance, should not be stored for more than two months, and uncooked biscuit dough for more than one month.

A basic biscuit mix can be flavoured in various ways, and each flavour stored in a Swiss roll shape, from which you cut slices of just the right thickness and bake them almost while the kettle boils. It would be worthwhile to double the quantities of the basic mix when making either, and cook a batch for instant eating and a batch to store.

Certain cakes freeze well, and all the recipes given here are freezer-tested for their qualities in retaining a 'good as new' flavour and texture when thawed after a month or more of storage.

If your home baked cakes appear, after freezing, to have a poor texture or flavour, this may be due to the choice of ingredients rather than to a fault in baking. Freshly bought flour should always be used as stale flour tends to deteriorate rapidly after freezing. Another factor which helps to retain the fresh flavour of the cake is the freshness of the fat used. Fats such as butter and margarine easily pick up cross-flavours from other foods if stored in the larder or refrigerator or simply left on an open dish in the kitchen. Eggs used in cakes for freezing ought always to be perfectly fresh and very well beaten. Again, they must be extra fresh to avoid that disappointing 'flatness' of flavour when the cake is thawed out. The beating must be thorough because egg whites

freeze more quickly than yolks and traces of unblended egg white will give the cake an uneven texture after thawing.

If you have time, make up a few fancy sandwiches as well as the standard meal size kind, and these will add an elegant touch to a tea-party table, or will add grace to a buffet spread. They store well for several weeks.

Here's how to make dainty pinwheel sandwiches for tea parties. Cut a sandwich loaf, lengthways, into thin slices and remove the crusts. Spread each slice with different coloured suitable fillings and roll up tightly as for a Swiss roll, but *lengthways*. Give the first roll a sharp press into shape; this is needed to make the final neat circle of each sandwich.

To freeze: Wrap the rolls individually in waxed paper and slide them into a polythene bag, excluding as much air as possible. Fasten the end securely. These pinwheel sandwiches are now ready for a party, or (if frozen) for any time you may need them during the next few weeks.

To prepare for serving: Remove the pinwheels from the freezer an hour or two before they are to be cut. Unwrap each roll, and place it join side down on a bread board. Cut into $\frac{1}{4}$-inch ($\frac{1}{2}$-cm.) slices with a really sharp knife.

Sandwich fillings
Cream cheese with sweet pickles
Cream cheese with chopped chicken and grated lemon rind
Cream cheese with chopped salami sausage
Cream cheese and chopped dates
Peanut butter, cream cheese and chives
Blue cheese, chopped bacon and chilli sauce
Cheddar cheese, grated, with chopped olives
Cream cheese and chopped olives
Sliced ham, tongue or corned beef with finely minced onion

Cream cheese and liver pâté with Worcestershire
    sauce
Chopped chicken, ham and almonds
Chopped cooked liver, crisp bacon and pickles
Chopped frankfurters and chilli sauce
Chopped prawns, cream cheese and lemon juice
Chopped prawns; cottage cheese and sweet pickles

Flavoured butters for pinwheel sandwiches

As these sandwiches look much prettier if the
filling contrasts well in colour with the bread, the
best method is to blend the filling with the butter
for spreading so that you can see what the colour
of the finished sandwiches will be.

Green butter: Cream 4 oz. (100 g.) butter, work
in 1 heaped tablespoon of very finely chopped
parsley, the juice of ½ lemon, salt and pepper to
taste.

Anchovy butter: Cream 4 oz. (100 g.) butter,
work in 4 chopped and pounded anchovy fillets, a
pinch of pepper and a few drops of pink food
colouring.

Mustard butter: Cream 4 oz. (100 g.) butter,
work in a level dessertspoon of French mustard.

Shrimp butter: Cream 2 oz. (50 g.) butter with a
small carton of potted shrimps (including their
own butter).

Ginger Danish pastries

½ oz./15 g. yeast

1¾ oz./35 g. castor sugar

8–12 tablespoons tepid milk

12 oz./350 g. plain flour

pinch salt

7 oz./200 g. butter

Filling A:

2 oz./50 g. ground almonds mixed with

    5 pieces chopped stem ginger and

    2 tablespoons ginger syrup

Filling B:

2 oz./50 g. ground almonds

5 pieces stem ginger, chopped

Cream the yeast with 1 teaspoon sugar. Pour
¼ pint (1½ dl.) of the tepid milk into the yeast.
Sift the flour and salt into a bowl, add the remaining sugar. Make a well in the centre and pour in
the yeast mixture. Work into the flour, adding the
remaining milk if necessary, to form a light but
dry dough. Knead until smooth.

Roll into an oblong ½ inch (1 cm.) thick and
place knobs of butter, the size of walnuts, over
two-thirds of the surface. Fold and roll as for
rough puff pastry and repeat once. Wrap in
greaseproof paper and place in a refrigerator or
cool place for 15 minutes, or until firm. Repeat
the rolling and folding until all the butter is
worked in. Chill for 1 hour or overnight. When
firm, roll out into a square ¼ inch (½ cm.) thick.
Cut in half.

Spread one half with the ground almond filling
A. Roll up into a sausage shape and cut into six
pieces. Cut each slice twice, two-thirds of the
way down, and fan out. Cut the other half into
squares. Fill half the squares with chopped ginger
and nuts (filling B) and fold the corners to the
centre. Make the rest of the pastry and filling into
your favourite shapes. Place on greased baking
sheets, covered, in a warm place to prove for
10–15 minutes. When the pastries have increased
their size by half, glaze with beaten egg white.
Place in a hot oven (450°F., 230°C., Gas Mark 8)
for 15–20 minutes until golden brown. Cool on a
wire tray.

To freeze: Seal tightly in polythene bags.

To prepare for serving: Thaw at room temperature allowing approximately 2 hours. *Makes
about 15 pastries.*

Ginger Danish pastries

Pork pie for high tea

For the pastry:
4 oz./100 g. plain flour
½ teaspoon salt
1½ oz./40 g. white fat or lard
2 tablespoons water
For the filling:
12 oz./350 g. pork sausage meat
pinch powdered sage
salt and pepper to taste

Combine the sausage meat, sage, salt and pepper, then mix thoroughly to make sure the seasonings are well mingled and set aside until the pastry has been made.

Sieve the flour and salt into a warm bowl. Measure the fat and water into a small saucepan and bring to a rolling boil. Pour immediately into the centre of the flour and mix to a smooth ball with a wooden spoon. Turn on to a board and knead until smooth. Reserving a little for the lid and decoration, shape the remainder into a round. Flatten the base and begin to shape the sides. Place the sausage meat in the centre and gradually work the pastry up round the sides to form an edge. Roll the reserved pieces out to form a lid. Damp the edge of the pie with a little beaten egg and milk and place the lid over the top. Pinch the edges to seal and snip round with scissors. Flute the edges of the pie. Make a hole in the centre and decorate with leaves made from any remaining pastry. Brush with beaten egg and fix a greased band of paper around the pie. Place the pie in the centre of a very hot oven (425°F., 220°C., Gas Mark 7) for 20 minutes, then lower the heat to 350°F., 150°C., Gas Mark 4 and bake for a further 40 minutes. After 30 minutes remove the band of paper and brush the whole pie with beaten egg.

To freeze: Allow to become completely cold. Wrap in a double thickness of foil and seal well.

To prepare for serving: Thaw at room temperature allowing at least 8 hours. *Serves 6.*

Basic biscuit dough

1 lb./450 g. plain flour and
  1 teaspoon baking powder
8 oz./225 g. butter or margarine
8 oz./225 g. castor sugar
2 eggs, beaten
½ teaspoon vanilla essence
little milk
½ teaspoon coffee essence

Sieve the flour and baking powder on to a piece of paper. Cream together the butter and sugar until light and fluffy. Add the beaten eggs, a little at a time, alternately with a few spoonfuls of sifted flour. Mix in the rest of the flour. Divide the mixture in half. Add the vanilla essence to one half and work it into the dough with sufficient milk to make a firm paste. Add the coffee flavouring to the other half of the mixture and work into the dough in the same way.

To freeze: Form each portion into a roll, and wrap in polythene tissue paper or aluminium foil. Seal and freeze.

To prepare for cooking: Thaw at room temperature until the roll is soft enough to slice into ¼-inch (½-cm.) slices and bake in a moderately hot oven (375°F., 190°C., Gas Mark 5) for 10–15 minutes. If liked decorate by lightly pressing a glacé cherry or almond in the centre.

One-stage peanut cookies

6 oz./175 g. plain flour
½ teaspoon baking powder
¾ teaspoon bicarbonate of soda
¼ teaspoon salt
4 oz./100 g. luxury margarine
3 tablespoons peanut butter
3 oz./75 g. castor sugar
3 oz./75 g. soft brown sugar
1 standard egg

Sieve the flour, baking powder, bicarbonate of soda and salt into a mixing bowl. Place the remaining ingredients in the bowl and beat together with a wooden spoon until well mixed (approximately 2–3 minutes). Gather the mixture together with the fingertips and form into a roll.

To freeze: Cut the roll into four pieces and place dividing papers between them. Wrap the rolls in paper or aluminium foil. If wrapped in polythene tissue paper, seal before freezing.

To prepare for cooking: Unwrap and thaw at room temperature until the dough is soft enough to slice. Cut thinly and place fairly far apart on a greased baking sheet. Bake in the centre of a moderately hot oven (375° F., 190°C., Gas Mark 5) for 7–10 minutes. Cool on a wire tray and then dip the cookies in melted chocolate. *Makes 30.*

One-stage treacle crisps

6 oz./175 g. plain flour
½ teaspoon cream of tartar
½ teaspoon bicarbonate of soda
¼ teaspoon salt
4 oz./100 g. luxury margarine
6 oz./175 g. castor sugar
3 tablespoons black treacle
2 oz./50 g. raisins, chopped

Sieve the flour, cream of tartar, bicarbonate of soda and salt into a mixing bowl. Place the remaining ingredients in the bowl and beat together with a wooden spoon until well mixed (approximately 2–3 minutes). Gather the mixture together with the fingertips and form into a roll.

To freeze: Cut the roll into four pieces and place dividing papers between them. Wrap the rolls in polythene tissue paper or aluminium foil. If wrapped in polythene tissue paper, seal before freezing.

To prepare for cooking: Unwrap and thaw at room temperature until the dough is soft enough to slice. Cut thinly and place the cookies fairly wide apart on a greased baking sheet. Bake in the centre of a moderately hot oven (375°F., 190°C., Gas Mark 5) for 7–10 minutes. Cool on a wire tray. *Makes 36–40.*

Date and nut loaf

1 lb./450 g. plain flour
1 teaspoon salt
2 teaspoons baking powder
4 oz./100 g. butter or margarine
4 oz./100 g. soft brown sugar
6 oz./175 g. cooking dates, chopped
2 oz./50 g. walnuts, chopped
1 egg
scant ½ pint/5 dl. milk

Sieve the flour, salt and baking powder into a mixing bowl. Rub the butter into the flour, add the sugar, dates and walnuts. Make a well in the centre, add the egg and sufficient milk to make a soft dough. Turn the dough on to a floured board and knead lightly, shape into a roll and place in a greased 2-lb. (1-kg.) loaf tin. Bake in the centre of a moderately hot oven (375°F., 190°C., Gas Mark 5) for 50–60 minutes.

To freeze: Cool completely and wrap in polythene tissue paper, double or heavy duty foil or thick polythene. Seal and freeze.

To prepare for serving: Thaw, still wrapped, at room temperature allowing approximately 6 hours. *Makes a 2-lb. (1-kg.) loaf.*

Banana fruit loaf

8 oz./225 g. self-raising flour
½ teaspoon salt
pinch mixed spice
2 oz./50 g. glacé cherries, washed and quartered
4 oz./100 g. castor sugar
1 heaped tablespoon golden syrup
2 oz./50 g. peel, chopped
2 oz./50 g. walnuts, coarsely chopped
4 oz./100 g. luxury margarine, sliced
1 lb./450 g. bananas, mashed
2 eggs

Sieve the flour, salt and spice into a mixing bowl. Dry the cherries and add to the bowl with the remaining ingredients then beat with a wooden spoon until evenly mixed. Spread in a 2-lb. (1-kg.) greased loaf tin, and bake in the centre of a moderate oven (350°F., 180°C., Gas Mark 4) for 1½–1¾ hours.

To freeze: Cool completely. Wrap in polythene tissue paper, a double thickness of foil, or thick polythene. Seal and freeze.

To prepare for serving: Thaw, still wrapped, at room temperature allowing approximately 6 hours. *Makes a 2-lb. (1-kg.) loaf.*

Gingerbread

Gingerbread

4 oz./100 g. margarine

6 oz./175 g. black treacle

2 oz./50 g. golden syrup

¼ pint/1½ dl. milk

2 eggs

8 oz./225 g. plain flour

2 oz./50 g. brown sugar

1 rounded teaspoon mixed spice

1 teaspoon bicarbonate of soda

2 teaspoons ground ginger

4 oz./100 g. sultanas (optional)

Grease and line a 7-inch (18-cm.) square cake tin. Warm together the margarine, treacle and syrup. Add the milk and cool. Beat the eggs and blend with the mixture. Sieve the dry ingredients together, add the cooled mixture and blend with a tablespoon. Add the fruit if required. Turn into the prepared tin. Bake in the centre of a slow oven (300°F., 150°C., Gas Mark 2) for 1¼–1½ hours.

To freeze: Turn out of the tin and cool on a wire tray. Wrap in a polythene bag with all air excluded. Tightly seal and freeze.

To prepare for serving: Thaw, sealed, at room temperature for 4–6 hours. Top with water icing and chopped almonds. *Makes a 7-inch (18-cm.) square cake.*

Plum cake ring

6 oz./175 g. butter or margarine

6 oz./175 g. soft brown sugar

3 large eggs

8 oz./225 g. self-raising flour

1 teaspoon baking powder

1 teaspoon mixed spice

8 oz./225 g. sultanas

6 oz./175 g. currants

4 oz./100 g. mixed cut peel

2 oz./50 g. glacé cherries

4 oz./100 g. walnuts, chopped

1 miniature bottle rum

To decorate:

crystallised fruits

castor sugar

Cream the butter and sugar until light and fluffy. Lightly whisk the eggs and beat into the butter and sugar mixture gradually. Sieve the flour, baking powder and mixed spice; fold into the creamed ingredients. Stir in the remaining ingredients and spoon the mixture into a greased and floured 3-pint (1¾-litre) ring mould. Or grease and flour the outside of a 1-lb. (454-g.) cocoa tin, weight it, and place centrally in a prepared 9-inch (23-cm.) cake tin. Bake in a hot oven (400°F., 200°C., Gas Mark 6) for 15 minutes then reduce the temperature to 300°F., 150°C., Gas Mark 2 for a further 2½–2¾ hours. Cool slightly before removing from the tin.

To freeze: Wrap tightly with polythene tissue paper and seal with freezer tape or wrap in a double thickness of foil. Store in a suitable container until frozen hard.

To prepare for serving: Thaw, wrapped, at room temperature. Pile crystallised fruits in the centre and sift with castor sugar. *Serves 18–20.*

Scones

1 lb./450 g. plain flour

1 teaspoon salt

2 teaspoons bicarbonate of soda and

    4 teaspoons cream of tartar, or

    8 teaspoons baking powder

3 oz./75 g. castor sugar

3 oz./75 g. butter or margarine

2 eggs, plus sufficient milk to make ½ pint/3 dl.

Sift the flour, salt, bicarbonate of soda and cream of tartar (or baking powder) together. Stir in the sugar. Rub in the fat. Beat the eggs and milk lightly together. Make a well in the centre of the flour mixture and pour the liquid into it. Stir lightly together to make a dough just firm enough to handle. Turn out on a floured board, roll or pat out by hand to ⅜-inch (¾-cm.) thickness. Cut into 2–2½-inch (5–6-cm.) rounds, bake on a floured baking sheet in a fairly hot oven (400°F., 200°C., Gas Mark 6) for 7–10 minutes. Cool.

To freeze: Pack into polythene bags or other suitable containers. Seal and freeze.

To prepare for serving: Thaw, covered, at room temperature for 2 hours. *Makes 16–20.*

Chocolate cake

8 oz./225 g. self-raising flour

2 oz./50 g. cocoa powder

6 oz./175 g. castor sugar

½ teaspoon salt

4 oz./100 g. luxury margarine

¼ pint plus 5 tablespoons/2 dl. milk

3 eggs

1½ tablespoons treacle

Grease an 8-inch (20-cm.) square or round cake tin. Line the bottom with greased paper. Sieve the dry ingredients into a large mixing bowl. Add the remaining ingredients and beat until well combined. Pour into the prepared tin. Bake slowly on the middle shelf of the oven at 300°F., 150°C., Gas Mark 2 for approximately 1¼–1½ hours. Cool in the tin for 5 minutes then turn out.

To freeze: Cool and freeze in a polythene bag with all air excluded. Seal tightly and freeze.

To prepare for serving: Allow to thaw at room temperature for 4–6 hours. *Makes an 8-inch (20-cm.) square or round cake.*

Feather sponge

5 oz./150 g. plain flour

1 oz./25 g. cornflour

2 teaspoons baking powder

½ teaspoon salt

5 oz./150 g. castor sugar

2 eggs, separated

6 tablespoons corn oil

6 tablespoons water

Line the bottom of two 7-inch (18-cm.) sandwich tins with greaseproof paper and grease lightly. Sieve the dry ingredients into a bowl. Mix together lightly with a fork the egg yolks, corn oil and water. Stir this into the dry ingredients. Whisk the egg whites until stiff, fold lightly into the mixture. Turn the mixture into the prepared tins. Bake in a moderately hot oven (375°F., 190°C., Gas Mark 5) for 25–30 minutes. Turn out and cool. Sandwich with jam.

To freeze: Wrap in a polythene bag with all the air excluded. Tightly seal and freeze.

To prepare for serving: Thaw, wrapped, at room temperature for 4–6 hours. *Makes a 7-inch (18-cm.) cake.*

Lemon curd

2 lemons

8 oz./225 g. castor sugar

3 oz./75 g. butter

3 eggs

Finely grate the zest from the lemons and squeeze the juice. Place the juice, zest and sugar in a double boiler, or basin over hot water, and stir until the sugar has dissolved. Add the butter and allow to melt. Remove from the heat, cool and beat in the eggs. Return to the heat and stir constantly without boiling until smooth and thick.

To freeze: Cool and pack into plastic containers or wide-necked jars. Seal and freeze.

To prepare for serving: Thaw for 30 minutes at room temperature.

Chapter Eleven

Hostess menus

Dinner parties these days tend to be more the informal 'come in for an evening meal' type than the grand affairs they used to be. Since cooking has become a delightful hobby to so many people, a small dinner for four or six at the most gives many a dedicated cook the chance to display his or her talents to an appreciative audience. This makes it much more fun to give such a party, than if one looks on it purely as a social obligation.

There was a time when people were brought up to consider it vulgar to talk enthusiastically about food, or even notice what they were eating as a guest in someone else's home. Nowadays people will freely discuss the menu and compliment a hostess on her cooking as well as exchanging gastronomic reminiscences and treasured recipes. In fact, the subject of food will often take precedence over the traditional rather dull dinner party conversation.

My selection of menus and suitable dishes, therefore, avoids the more obvious choices, such as prawn cocktail, which figure in every hostess's repertoire. The quantities are for six people, and most of the dishes are unusual enough to make an agreeable talking point during the meal, and afterwards.

The three-course dinner is now generally accepted as suitable for even formal entertaining at home (your husband's wealthy relatives; his boss; in fact anyone you don't know well and would like to impress). But just because the freezer makes it so easy, I have included one four-course menu, in case you feel like serving a hot savoury, just for a change. Any of the other menus can be extended to four courses by serving a tempting cheese platter (some English and some foreign cheeses, please) or a beautiful bowl of fresh fruit as dessert if the meal has been rather a substantial one.

To impress your guests the fruit bowl should contain one rather unusual or expensive fruit. A pineapple as a centrepiece, for instance, looks extremely attractive, although at times these are quite cheap; Chinese gooseberries or lychees in their mysterious prickly brown overcoats are now sold in most large towns in season, or the gorgeous crimson or orange-skinned plums and nectarines which are now accepted arrivals on the midwinter scene. For the rest, add crisp, juicy apples, pears and oranges, and ripe but still firm bananas to round out the selection.

Menus for special dinner parties

The following three menus are planned to give perfect balance to each meal. Reap the rewards of your labours and surprise your guests by serving luxury foods out of season.

Menu 1
Rich chicken liver pâté
Individual quiches
Apricot dessert

Menu 2
Gazpacho
Chicken with lemon cream
French apple pie

Menu 3
Silberbissen
Coquilles St. Jacques
Strawberries Romanoff
Délices de fromage

MENU 1

Rich chicken liver pâté

1 small onion, finely chopped

2 tablespoons corn oil

4–6 oz./100–175 g. fresh chicken livers

1 dessertspoon flour

1 hard-boiled egg yolk, mashed

6 oz./175 g. butter, softened

1 tablespoon sherry

2 tablespoons double cream

salt and pepper to taste

1 clove garlic, chopped (optional)

Fry the onion in the oil until golden, and remove from the pan. Trim the livers; wash, dry and coat in the flour. Put the livers in the pan, cover and cook gently for 5–6 minutes. When lightly cooked, chop roughly and pass them, with the onions, through a sieve. Add the hard-boiled egg yolk, butter, sherry, cream and seasoning. Mix together thoroughly until completely smooth and well blended or put into an electric blender.

To freeze: Pack in a foil pudding basin. Seal with a double thickness of foil and freeze.

To prepare for serving: Thaw at room temperature allowing 3–4 hours. Serve with toast. *Serves 6.*

Individual quiches

6 oz./175 g. shortcrust pastry

1 tablespoon corn oil

1 small onion, finely chopped

2 rashers streaky bacon, finely chopped

1 egg yolk

½ pint/3 dl. milk

1 tablespoon parsley, chopped

½ teaspoon salt

pinch pepper

1 oz./25 g. cheese, grated

Roll out the pastry and line six 3-inch (7½-cm.) patty tins. Bake blind with baking beans in a fairly hot oven (400°F., 200°C., Gas Mark 6) for 5 minutes. Remove the baking beans and bake for a further 5 minutes. Heat the corn oil. Add the onion and bacon and fry without browning.

Remove from the pan and drain.

Whisk the egg yolk and milk. Stir in the onion, bacon, parsley, salt, pepper and cheese. Divide the mixture between the six pastry cases. Bake in a moderately hot oven (375°F., 190°C., Gas Mark 5) for 15 minutes, or until the filling is set. Cool on a wire tray.

To freeze: When cold, freeze in a large plastic container until solid, then arrange in a plastic or foil container. Return to the freezer.

To prepare for serving: Thaw covered, at room temperature and, if liked, warm gently before serving. Garnish each quiche with a slice of tomato. *Serves 6.*

Apricot dessert

1 lb./450 g. apricots

¼ pint/1½ dl. water

3 oz./75 g. castor sugar

little milk

1½ oz./40 g. cornflour

2 eggs, separated

¼ pint/1½ dl. single cream

1 dessertspoon apricot brandy (optional)

Halve and stone the apricots. Place in a saucepan with the water and sugar. Cover and cook slowly for 15 minutes or until tender. Rub the apricots through a sieve. Make up the purée to 1 pint (6 dl.) with milk. Blend the cornflour and egg yolks with a little of the purée. Heat the remainder. Stir in the blended cornflour and bring to the boil, stirring. Cook for a further 3 minutes, stirring all the time to prevent lumps from forming. Pour into a large mixing bowl and leave to cool. Stir in the cream and apricot brandy, if used. Whisk the egg whites until stiff and fold lightly into the apricot mixture. Pour into a dampened 1-lb. (450-g.) loaf tin and chill.

To freeze: Cover closely with double or heavy duty foil. Freeze.

To prepare for serving: Thaw, uncovered, at room temperature, allowing approximately 6 hours. Turn out of the tin and decorate with fresh fruit and whipped cream. *Serves 6.*

MENU 2
Gazpacho

$\frac{1}{4}$ pint/1$\frac{1}{2}$ dl. olive oil
3 cloves garlic, crushed
1 small onion, very finely chopped
1 teaspoon salt
5 tablespoons vinegar
1 large ripe avocado pear, peeled and stoned
1 15-oz./425-g. can tomatoes
1 15-oz./425-g. can tomato juice

Beat thoroughly with a rotary beater, or liquidise in a blender, the oil, garlic, onion, salt and vinegar. Dice the flesh from the avocado pear. Beat or liquidise the canned tomatoes, tomato juice and avocado pear. Press through a sieve. Blend the two mixtures together, in two batches if necessary.

To freeze: Pour into a polythene bag or plastic container. Seal, leaving a 1-inch (2$\frac{1}{2}$-cm.) head-space, and freeze.

To prepare for serving: Thaw overnight in the refrigerator or at room temperature until liquid but still chilled. Stir well and adjust seasoning. Serve sprinkled with fresh chopped parsley and chives. *Serves 6.*

Chicken with lemon cream

1$\frac{1}{2}$ oz./40 g. butter
6 chicken breasts
2 tablespoons dry sherry
3 tablespoons dry white wine
grated zest of 1 lemon
1$\frac{1}{2}$ tablespoons lemon juice
salt and pepper to taste
$\frac{1}{4}$ pint/1$\frac{1}{2}$ dl. single cream
$\frac{1}{4}$ pint/1$\frac{1}{2}$ dl. double cream

Melt the butter in a frying pan, add the chicken pieces and fry until browned all over (approximately 5 minutes). Transfer the chicken to a shallow baking tin. Add the sherry, white wine, grated lemon zest, lemon juice and seasoning to taste to the juices in the pan. Stir well over gentle heat until thoroughly blended. Remove from the heat, add the cream, stir again and pour over the chicken.

124

To freeze: Cool rapidly, cover and seal the dish with a double thickness of foil before freezing.

To prepare for serving: Thaw at room temperature and transfer to an ovenproof serving dish. Sprinkle with 2 oz. (50 g.) grated Gruyère cheese, put in a moderate oven (350°F., 180°C., Gas Mark 4) for 35 minutes. Brown under the grill. Place a lemon butterfly on each portion, sprinkle with chopped parsley and serve with fluffy long-grain rice. *Serves 6.*

French apple tart

French apple tart

1 oz./25 g. butter
1 teaspoon cinnamon
4 oz./100 g. soft brown sugar
4 large cooking apples
For the pastry crust:
5 oz./150 g. plain flour
$\frac{1}{2}$ teaspoon salt
4 oz./100 g. butter
1 egg, lightly beaten
2 tablespoons cold water

Butter a 9-inch (23-cm.) sandwich tin and line the bottom with a circle of buttered greaseproof paper. Sprinkle with cinnamon and brown sugar. Peel and core the apples and slice as thinly as

Butter a 9-inch (23-cm.) sandwich tin and line with greaseproof paper.

Sprinkle the base with cinnamon and brown sugar. Cover the bottom of the dish with concentric circles of apple slices.

Fill the tin with more layers of apple slices, well pressed down. Lightly press the circle of pastry on top.

Wrap in double or heavy duty foil and overwrap with polythene.

possible. Cover the bottom of the dish with concentric circles of apple slices, pressing them down firmly. Fill the tin with more layers of apple slices, well pressed down. Top with the rest of the butter.

Make the pastry crust, mixing the ingredients with a fork as the dough will be very soft. Chill before rolling out to make it easier to handle. Roll out to a circle large enough to fit the dish and press lightly on top.

To freeze: Wrap in a double thickness of foil or sheet polythene. Seal and freeze.

To prepare for serving: Put the pie in a very hot oven (450°F., 230°C., Gas Mark 8) while still frozen. Immediately reduce heat to 375°F., 190°C., Gas Mark 5. Bake for 30–35 minutes, remove from the oven and immediately invert on a serving dish and remove the greaseproof paper. Serve warm if liked with sweetened whipped cream. *Serves 6*.

MENU 3
Silberbissen

1 large grapefruit, halved

12 oz./350 g. cream cheese

3 tablespoons soured cream

1 tablespoon onion, finely chopped

salt and pepper to taste

1 tablespoon lemon juice

2 tablespoons walnut halves, chopped

Remove the grapefruit segments. Beat together the cream cheese, soured cream, onion, salt and pepper and lemon juice. Stir in lightly the grapefruit segments and chopped walnuts.

To freeze: Spoon into plastic containers. Seal, leaving a ½-inch (1-cm.) headspace, and freeze.

To prepare for serving: Thaw at room temperature and spoon into cocktail glasses lined with shredded lettuce. Serve with fingers of toast. *Serves 6*.

Coquilles St. Jacques

¼ pint/1½ dl. dry white vermouth or
   ¼ pint plus 4 tablespoons/2 dl. dry white wine
salt and pepper to taste
1 bay leaf
2 tablespoons finely chopped onion
1 lb./450 g. scallops
8 oz./225 g. mushrooms, sliced
1½ oz./40 g. butter
1½ oz./40 g. flour
¼ pint/1½ dl. milk
2 egg yolks
¼ pint/1½ dl. double cream
1 teaspoon lemon juice

Simmer together for 5 minutes the vermouth or wine, salt, pepper, bay leaf and onion. Add the scallops and sliced mushrooms. Add sufficient water just to cover, simmer covered for 5 minutes. Remove the scallops and mushrooms and boil the stock quickly to reduce to ½ pint (3 dl.). Melt the butter, stir in the flour, blend in the stock and milk. Cook, stirring, for 1 minute. Remove from the heat. Blend the egg yolks with the cream, gradually stir the hot sauce into the egg and cream mixture. Return to the pan and reheat for 2 minutes but do not boil, stirring all the time. Season to taste with salt, pepper and lemon juice. Cut the scallops into bite-size pieces, stir with the mushrooms into the sauce. Butter six deep scallop shells, fill with the mixture.

To freeze: Cool rapidly. Cover closely with double or heavy duty foil and freeze.

To prepare for serving: Thaw in the refrigerator for about 4 hours, sprinkle with a little grated Gruyère cheese, dot with butter and grill for 15 minutes or until well browned. *Serves 6.*

Strawberries Romanoff

1½ lb./675 g. frozen strawberries (sweetened)
¼ pint/1½ dl. orange juice
7 tablespoons Curaçao or Cointreau
½ pint/3 dl. double cream
1 oz./25 g. icing sugar, sifted

Partially thaw the strawberries. Combine the orange juice and liqueur. Pour over the straw-

berries. Keep chilled. Place in a glass serving dish. Whip the cream and icing sugar and pile on top. *Serves 6.*

Délices de fromage

4 slices streaky bacon, trimmed
3 oz./75 g. Gruyère cheese, grated
1 oz./25 g. Parmesan cheese, grated
1 small onion, grated
1 teaspoon mayonnaise
½ teaspoon dry mustard
6 large slices white bread

Grill the bacon until crisp and crush finely. Mix the grated cheese, onion and bacon with the mayonnaise and mustard. Toast the bread on one side. Trim, halve, spread the mixture on the un-toasted side.

To freeze: Stack on a double thickness of foil, dividing each slice with foil. Wrap, seal and freeze.

To prepare for serving: Thaw, wrapped, at room temperature. Unwrap, then grill, cheese side up, until golden. *Serves 6.*

MORE PARTY DISHES
Consommé

8–12 oz./225–350 g. shin of beef
2 carrots, sliced
1 leek, sliced
1 stick celery, sliced
2 pints/generous litre brown stock
4 stalks parsley
1 sprig thyme
½ bay leaf
3–4 egg whites, lightly whisked

Trim the fat from the meat and cut into fine strips. Put it into a saucepan with the vegetables and stock. Tie the parsley stalks, thyme and bay leaf with thread and add to the ingredients in the pan, together with the egg whites. Bring slowly to simmering point, stirring all the time. When the liquid begins to cloud, stop stirring and continue to simmer gently for 10–15 minutes, or until the egg has hardened and the consommé is clear and brilliant. Put a clean, wet cloth over a large mixing bowl and ladle the consommé through it. Do not press the soup through the cloth. Discard the meat and vegetables.

To freeze: Pack in plastic containers. Seal and freeze.

To prepare for serving: Immerse the container in hot water. Gently reheat the contents in a saucepan. Add 3 tablespoons dry sherry and correct the seasoning before serving. *Serves 6.*

French onion soup

1 oz./25 g. margarine
1 lb./450 g. onions, skinned and sliced
1½ pints/scant litre stock or 2 stock cubes and water
salt to taste
1 oz./25 g. cornflour
¼ pint/1½ dl. cold water

Melt the margarine in a pan. Add the sliced onions, cover and cook gently for 20–30 minutes until soft but not coloured. Remove the lid and allow to cook until brown. Stir in the stock and salt, re-cover and simmer for 30 minutes. Blend the cornflour and water together, stir into the soup. Continue to cook, stirring, until thick.

To freeze: Cool and pour into plastic or other suitable containers. Seal and freeze.

To prepare for serving: Gently reheat in a pan. Ladle into soup plates. Serve with slices of toasted French bread and grated Parmesan cheese. *Serves 6.*

Cream of cucumber soup

2 young cucumbers
1 oz./25 g. butter
salt and pepper to taste
pinch sugar
¼ pint/1½ dl. chicken stock
¼ pint/1½ dl. white sauce
green colouring

Wipe and peel the cucumbers, slice in half lengthways, remove the seeds. Cut into thick slices, blanch in boiling water for 30 seconds and drain.

Melt the butter in a heavy saucepan, add the cucumbers, seasoning and sugar. Cover and cook gently until soft. Add the chicken stock and white sauce. Bring to the boil. Pass through a fine sieve, or liquidise in a blender, and return to the saucepan. Thin down with a little more stock if necessary. Check the seasoning and add a little green colouring.

To freeze: Cool and pack into a plastic container, leaving a 1-inch (2½-cm.) headspace.

To prepare for serving: Immerse the container in hot water to loosen the contents. Reheat gently in a saucepan. Add a 2½-oz. (71-g.) carton single cream and reheat without boiling. *Serves 6.*

Vichyssoise

6 large leeks
4 medium potatoes
2 oz./50 g. butter
1½ pints/scant litre chicken stock or
    2 stock cubes and water
salt and pepper to taste
To garnish:
1 heaped tablespoon chopped chives

Trim all green parts from the leeks, wash thoroughly and cut into short lengths. Peel and slice the potatoes. Heat the butter in a saucepan and sauté the leeks gently until soft but not coloured, then add the potatoes, chicken stock and seasoning and simmer until the vegetables are tender. Press through a fine sieve or liquidise in an electric blender.

To freeze: Pack in plastic containers, leaving a 1-inch (2½-cm.) headspace. Cover and freeze.

To prepare for serving: Thaw at room temperature, stir in ¼ pint (1½ dl.) fresh or soured cream. Sprinkle each serving with chopped chives. *Serves 6.*

Piquant avocado dip

2 medium avocados
2 tablespoons lemon juice
6 oz./175 g. cream cheese
1 teaspoon salt
¼ teaspoon pepper
dash Tabasco sauce
1 tablespoon finely grated onion

Stone the avocados and mash or sieve the flesh until smooth. Add the remaining ingredients and mix well.

To freeze: Pack into a suitable plastic container. Seal and label.

To prepare for serving: De-frost in the refrigerator overnight, stir thoroughly and serve chilled. If required quickly, hold the container under warm water until the dip can be broken up and mixed with a fork until smooth. Garnish with lemon and sprigs of parsley. Serve with potato crisps and small savoury biscuits. *Serves 6.*

Terrine of pork

Terrine of pork

1 lb./450 g. lean pork, minced
8 oz./225 g. pork sausage meat
4 oz./100 g. rolled oats
rind and juice of ½ lemon
salt and pepper to taste
½ teaspoon sage
1 onion, grated
1 egg, beaten
6 oz./175 g. streaky bacon rashers

Mix together all the ingredients except the rashers. De-rind the rashers and place on a board, then stretch them by stroking lengthways gently, using a dinner knife. Arrange in a 2-lb. (1-kg.) loaf tin. Carefully press the meat mixture into the tin and level off the top. Cover with foil and put in a baking tin of water. Cook for 1½ hours in a moderate oven (350°F., 180°C., Gas Mark 4). When cooked, place a weight on top and leave overnight.

To freeze: Ease the pâté out of the loaf tin. Cut into 12 slices. Put polythene tissue paper or foil dividers between every slice; wrap closely in double or heavy duty foil. Freeze.

To prepare for serving: Open the package and spread out the portions. Thaw at room temperature for about 1 hour. *Serves 6.*

Glazed salmon trout

Glazed salmon trout

1 2½–3-lb./1–1¼-kg. salmon trout, thawed
6 parsley stalks
½ lemon, cut into thick slices
To garnish:
2 tablespoons powdered gelatine, soaked in
    2 tablespoons water
2 pints/generous litre water
juice of ½ lemon
stuffed olives, sliced
1 canned red pimento, sliced
slices of lemon

Wrap the fish in a well sealed foil parcel with the parsley stalks and lemon. Place on a baking sheet and cook in the centre of a moderate oven (350°F., 180°C., Gas Mark 4) allowing 20 minutes per lb.

(450 g.). Unwrap and cool on a wire tray.

To garnish, heat a little of the water and pour on to the soaked gelatine; stir to dissolve before adding to the rest of the water with the lemon juice. Leave in a cool place until on the point of setting. Put a clean tray under the wire rack and spoon the jelly over the fish. Decorate with the olives and pimento. Pour the remaining jelly into an oblong tin and allow to set. Turn out of the tin, dice and arrange on the base of the serving dish. Transfer the fish to the serving dish and garnish with lemon slices. *Serves 6–8.*

Rice with mushrooms and scampi

Rice with mushrooms and scampi

8 oz./225 g. button mushrooms

10 oz./275 g. long-grain rice

3 oz./75 g. butter

12 oz./350 g. frozen scampi, thawed

1 tablespoon brandy

1 tablespoon tomato purée

2 tablespoons flour

¼ pint/1½ dl. milk

2 tablespoons oil

salt and pepper

3 tablespoons grated Parmesan cheese

Wash and dry the mushrooms. Cook the rice in boiling salted water until tender (from 14–20 minutes depending on the type used). In a saucepan heat 1 oz. (25 g.) of the butter, add the scampi and cook for 2 minutes. Add the brandy and when evaporated stir in the tomato purée and flour. Add the milk, stir and cook for another minute. Cover and simmer gently for 10 minutes. Meanwhile, in another pan, heat 1 oz. (25 g.) butter with 1 tablespoon oil and sauté the mushrooms gently for 3½ minutes; season with salt and pepper. When the rice is cooked, drain thoroughly and return to the saucepan with the remaining butter and the cheese. Heat gently until the butter and cheese combine with the rice.

To freeze: Cool the rice and pack separately in a foil pudding basin. Cover with double or heavy duty foil and freeze. Pack the scampi mixture into a carton. Seal and freeze.

To prepare for serving: Thaw at room temperature allowing approximately 4–5 hours. Put the rice in a sieve or colander and place over a saucepan of boiling water. Cover and heat through. Add a lightly beaten egg yolk and ¼ pint (1½ dl.) single cream to the scampi mixture. Reheat in a saucepan. Do not allow to boil. *Serves 6.*

Prawn curry

3 pints/1¾ litres fresh prawns or

  1 8-oz./226-g. packet frozen prawns

2 tablespoons corn oil

2 medium onions, chopped

3 tablespoons cornflour

2 chicken stock cubes

3 teaspoons curry paste

½ teaspoon salt

1½ tablespoons demerara sugar

1 14-oz./396-g. can tomatoes

1 bay leaf

1 pint/6 dl. water

Peel the prawns and remove the veins. (If using frozen prawns, de-frost.) Heat the corn oil. Add the onion and fry until soft without browning. Add the cornflour, stock cubes, curry paste, salt, sugar, tomatoes and bay leaf. Stir in the water and bring to the boil, stirring. Cover and simmer for 30 minutes. Add the prawns and simmer for a further 20 minutes.

To freeze: Cool, put into a suitable container, cover and seal before freezing.

To prepare for serving: Turn carefully into a saucepan and reheat over gentle heat. Serve with plain boiled rice. *Serves 6.*

Boeuf à la bourguignonne

2 lb./900 g. best stewing steak
2 tablespoons flour
salt and pepper
1½ oz./40 g. dripping or lard
4 oz./100 g. streaky bacon, diced
12 small onions or shallots
1 clove garlic, crushed
pinch thyme
½ bay leaf
1 15-oz./425-g. can cream of mushroom soup
2 oz./50 g. mushrooms, sliced
¼ pint/1½ dl. Burgundy

Trim the meat and cut into neat 1-inch (2½-cm.) squares, toss in the seasoned flour and fry in hot fat in a flameproof casserole for a few minutes to seal. Remove the meat from the casserole and keep hot. Fry the bacon, and return the meat to the casserole. Add six onions, the garlic, thyme, bay leaf, mushroom soup, mushrooms and Burgundy. Cook for 1½ hours in a warm oven (325°F., 170°C., Gas Mark 3). Add the remaining onions.

To freeze: Remove the bay leaf and cool rapidly. Pack in plastic or foil containers. Seal and freeze.

To prepare for serving: Thaw at room temperature allowing approximately 6 hours for thawing. Cook for 40 minutes in a warm oven (325°F., 170°C., Gas Mark 3). *Serves 6.*

Boeuf à la provençale

2 lb./900 g. chuck or buttock steak
4 oz./100 g. belly of pork
1 tablespoon olive oil
1 11-oz./312-g. can tomatoes
¼ pint/1½ dl. stock or water
6 parsley stalks and ½ bay leaf, tied with thread to make
   a bouquet garni
½ teaspoon salt
pinch pepper
½ teaspoon rosemary
¼ pint/1½ dl. dry white wine
6 green olives, stoned

Trim the beef and cut into 1-inch (2½-cm.) cubes. Trim the pork and dice finely. Heat the oil in a

Boeuf à la bourguignonne

large saucepan and add the pork; fry until the fat is rendered down. Add the beef to the pork and fry briskly to brown on all sides. Add the tomatoes, with the juice, the stock, bouquet garni, salt, pepper and rosemary. Cover with a lid and simmer very gently for 1 hour. Add the wine and continue to cook for a further 15 minutes.

To freeze: Add the olives and cool. Put into a plastic, foil or other suitable container, cover closely and seal before freezing.

To prepare for serving: Turn into a saucepan, adjust the seasoning and thicken with 1 oz. (25 g.) beurre manié. *Serves 6.*

Beurre manié: Work butter and flour together, using one-and-a-half parts of butter to one part flour. This can be made in bulk and stored in the refrigerator for up to 6 weeks or it can be stored in the freezer for up to 3 months. Weigh the required amount and stir into soups or stews to thicken.

Rich beef casserole

2 lb./900 g. chuck or buttock steak
1 oz./25 g. seasoned flour
2 tablespoons cooking oil
1 large onion, finely chopped
1 clove garlic
1 teaspoon salt
1 tablespoon tomato purée
1 teaspoon sugar
½ teaspoon dried basil
pinch pepper
½ pint/3 dl. stock or water
¼ pint/1½ dl. dry red wine

Cut the meat into 1-inch (2½-cm.) cubes and toss in the seasoned flour. Heat the oil in a large saucepan and when it is very hot, fry the meat to brown on all sides. Lower the heat and add the onion. Crush the garlic with the salt and add it to the ingredients in the pan. Continue to cook until the onion softens. Mix in the tomato purée and add the sugar, basil, pepper and stock. Cover with a lid and simmer over a gentle heat for 45 minutes before adding the wine; cook for a further 30 minutes.

To freeze: Cool rapidly, put into a suitable plastic container, cover closely and seal before freezing.

To prepare for serving: Turn into a saucepan and reheat gently.   *Serves 6.*

Boeuf stroganoff

Boeuf stroganoff

3 tablespoons cooking oil or 2 oz./50 g. butter

1 onion, peeled and sliced

1 small clove garlic, crushed

1¼ lb./575 g. stewing steak, cut into thin strips
    1½ inches/3½ cm. long

4 oz./100 g. mushrooms, sliced

pinch paprika

2 teaspoons Worcestershire sauce

1 packet tomato soup

¾ pint/4 dl. water

Melt the oil or butter in a saucepan and add the onion, garlic and meat. Cover, and simmer gently for 30 minutes. Add the mushrooms, paprika, Worcestershire sauce and contents of the packet of tomato soup. Gradually stir in the water, bring to the boil, cover and simmer for 30 minutes or until the meat is tender.

To freeze: Cool and spoon into a foil or plastic container. Seal and freeze.

To prepare for serving: Heat in a warm oven (325°F., 170°C., Gas Mark 3) and add ¼ pint (1½ dl.) yogurt or soured cream when the beef is hot but not bubbling.   *Serves 6.*

Rich beef stew with mustard croûtons

1½ lb./675 g. stewing steak

1 tablespoon corn oil

1 large onion

salt and pepper to taste

1 tablespoon French mustard

1 beef stock cube

2 oz./50 g. raisins

¾ pint/4 dl. light ale

½ oz./15 g. cornflour

For the mustard croûtons:

2 slices stale white bread

1 teaspoon dry mustard

oil for frying

chopped parsley to garnish

Cut the meat into even-sized dice. Heat the corn oil in a flameproof casserole and sauté the peeled and chopped onion and meat in it until brown. Add salt and pepper to taste, the French mustard, stock cube and raisins. Stir in the ale, bring to the boil, cover and simmer gently until the meat is tender, about 1½ hours. Moisten the cornflour with a little cold water, stir into the casserole, bring to the boil stirring all the time. Cook for a further 3 minutes. To make the croûtons, cut the bread into small triangles. Place in a shallow bowl and sift mustard over them through a sieve, shaking the bowl gently to coat the bread. Fry in very hot oil until golden brown on both sides. Drain well.

To freeze: Cool rapidly and pack the stew and croûtons separately in plastic containers.

To prepare for serving: Turn the stew into a saucepan and reheat gently. Garnish with croûtons, thawed at room temperature and sprinkled with parsley. *Serves 4.*

Gingered pork curry

Sauté de veau Marengo

2 lb./900 g. shoulder of veal

2 oz./50 g. butter

2 tablespoons oil

2 medium onions, peeled and chopped

1 tablespoon flour

¼ pint/1½ dl. dry white wine

1 tablespoon tomato purée

½ bay leaf, tied together with 4 parsley stalks and
   1 sprig thyme

¼ teaspoon basil

salt and pepper

½ teaspoon sugar

8 oz./225 g. button mushrooms, washed and trimmed

Cut the meat into 2-inch (5-cm.) cubes. Heat the butter and oil together in a large heavy frying pan; add the meat to the pan and brown over a brisk heat. Remove the meat and sauté the onions until soft but not coloured. Sprinkle the flour over the onions and cook gently for 1 minute. Transfer the onion to a saucepan and add the meat to it. Pour the wine over the meat; add the tomato purée, bouquet garni, basil, salt, pepper and sugar. Cover the saucepan with greaseproof paper and a tight-fitting lid. Simmer for 1 hour. Halve the mushrooms and add to the saucepan. Continue to cook for 3 minutes.

To freeze: Cool rapidly and spoon into plastic or other suitable containers, leaving a ½-inch (1-cm.) headspace. Seal and freeze.

To prepare for serving: Thaw at room temperature for about 6 hours. Reheat in a saucepan. Garnish with triangles of fried bread before serving. *Serves 6.*

Gingered pork curry

8 oz./225 g. desiccated coconut

1 pint/6 dl. boiling water

4 oz./100 g. butter

2 onions, peeled and chopped

2 green peppers, chopped

2 sticks celery, chopped

3 lb./1¼ kg. bladebone of pork, cubed

2 tablespoons flour

2–3 tablespoons curry powder

1 teaspoon ground ginger

2 tablespoons soy sauce

2 cooking apples, peeled and chopped

salt to taste

Place the coconut in a basin and pour over the boiling water. Allow to stand for at least 2 hours (overnight if possible) then strain off the 'milk' and squeeze the coconut to get out as much liquid as possible. Melt the butter and fry the onion, chopped pepper and celery until the vegetables are softened. Add the meat and toss over high heat until sealed. Sprinkle over the flour, curry powder and ginger, stir well and allow to cook for about 3 minutes. Gradually add the soy sauce, coconut milk and apple and bring to the boil, stirring constantly. Add a little extra water if too thick. Cover, simmer for 20 minutes and add salt if desired.

To freeze: Divide the curry between eight boilable bags if individual portions are required, or pack in larger quantities, press out excess air. Seal and freeze.

To prepare for serving: Place the required number of bags, still frozen, in a pan of boiling water and reheat for 15–25 minutes, depending on quantity of contents. Serve with fluffy boiled rice. (Portions of cooked rice can be frozen in boiling bags to be reheated in the same way as the curry.) *Serves 8.*

Pork satay with golden and white rice

2 lb./900 g. pork fillet

3 tablespoons peanut butter

1 teaspoon ground coriander

1 large onion, finely grated

1 teaspoon salt

1 tablespoon brown sugar

2 tablespoons lemon juice

1 tablespoon soy sauce

1 teaspoon Tabasco sauce

pinch ground ginger

Cut up the pork fillet into 1-inch (2½-cm.) cubes. Leave overnight in the refrigerator in a marinade made by mixing the remaining ingredients well together.

To freeze: Pack meat and marinade in container, seal, label.

To prepare for serving: De-frost, drain and reserve the marinade. Thread the meat, pieces of onion and a few bay leaves, if liked, on skewers or spread on foil in the grill pan. Grill under high heat for 5 minutes. Turn, brush with the marinade and grill for a further 5 minutes. Reduce the heat and continue basting and turning for 20 minutes, or until the meat is tender. Serve with golden and white rice. *Serves 6.*

Golden and white rice

Cook 4 oz. (100 g.) long-grain rice in boiling salted water. Cook a similar amount of rice in another pan, together with a few saffron strands (or a pinch of powdered saffron) and ½ teaspoon

turmeric. Refresh the white rice, shake dry and turn on to a serving dish. Refresh the golden rice, place on top of the white rice. Dissolve 1 oz. (25 g.) sugar in 2 fluid oz. (50 ml.) orange juice. Add 1 tablespoon very thin slivers of orange peel and boil in the syrup for 1 minute. Pour over the rice. *Serves 6.*

Stuffed fillet of pork

1 lb./450 g. pork fillet (tenderloin)

2 oz./50 g. butter or margarine

1 medium onion, peeled and finely chopped

4 oz./100 g. mushrooms, thinly sliced

1 oz./25 g. flour

½ pint/3 dl. chicken stock

pinch pepper

½ teaspoon salt

For the filling:

1 small onion, peeled and finely chopped

8 oz./225 g. streaky bacon, cooked and diced

3 tablespoons breadcrumbs

4 oz./100 g. sausage meat

¼ teaspoon salt

¼ teaspoon mixed herbs

1 egg, beaten

Trim the pork and cut into six pieces. Beat each piece between two sheets of wet greaseproof paper until very thin. Combine all the ingredients for the filling and spread equal portions on to each piece of pork. Fold the opposite sides to overlap, then tuck the ends over, parcel-fashion. Secure with cocktail sticks.

Melt the butter in a saucepan and cook the pork briskly until brown on all sides. Remove from the pan. Add the onion and mushrooms to the pan and cook gently until soft, but not coloured. Return the pork to the pan, add the flour, liquid and seasoning. Cover with a tight-fitting lid and simmer for 15 minutes.

To freeze: Cool rapidly. Pack into plastic or other suitable containers, leaving a ½-inch (1-cm.) headspace. Seal and freeze.

To prepare for serving: Immerse the container in hot water. Gently reheat in a saucepan. Serve with plain boiled rice. *Serves 6.*

New Zealand guard of honour

2 best ends of neck of lamb, 6 bones each, de-frosted
few cutlet frills
glacé cherries

This is a very professional looking roast, very easy to cook and serve. The two best ends of neck must each have the same number of cutlets, five, six or seven. Remove the chine bone from each joint, trim and clean the bone tips of 1½ inches (3½ cm.) of fat. Place the two joints together in a roasting pan, pressing the bones together to cross in the centre and protrude at each side. Pad the bone tips and roast in a well greased roasting pan in a moderately hot oven (375°F., 190°C., Gas Mark 5) for 1 hour. To serve, top the bone tips with cutlet frills and glacé cherries; garnish with parsley and serve with roast potatoes. *Serves 6.*

Sautéed sweetbreads

1½ lb./675 g. sweetbreads
¾ pint/4 dl. chicken stock or 1 stock cube and water
1½ oz./40 g. seasoned flour
beaten egg and breadcrumbs for coating

Soak the sweetbreads in salted, cold water for 1 hour. Drain and rinse. Put the sweetbreads in a saucepan, cover with water and bring to the boil. Drain and cool. Remove any fat. Simmer in the stock for 25 minutes or until tender. Drain, cool and cut into bite-size pieces. Toss in seasoned flour, dip in egg and coat with breadcrumbs.

To freeze: Pack in a polythene bag. Seal tightly and freeze.

To prepare for serving: Thaw, covered, in the refrigerator overnight. Fry in butter and serve with tomato sauce. *Serves 6.*

Chicken with pineapple

2 oz./50 g. butter or margarine
1 small green pepper, de-seeded and sliced
2 oz./50 g. flour
½ pint/3 dl. milk
½ pint/3 dl. chicken stock
1½ lb./675 g. cooked chicken, diced
1 12-oz./340-g. can pineapple cubes, drained
salt and pepper to taste

Heat the butter and fry the sliced pepper gently for 5 minutes. Set aside and stir the flour into the butter. Combine the milk and chicken stock and gradually add to the flour and butter. Stir over a moderate heat until the sauce is smooth and thick. Add the green pepper, chicken, pineapple, salt and pepper and a little of the fruit juice if liked.

To freeze: Cool quickly. Pack into polythene or other suitable containers, leaving a ½-inch (1-cm.) headspace.

To prepare for serving: Turn into a double boiler and heat through, for approximately 30 minutes. *Serves 6.*

Chicken paprika

1 3-lb./1¼-kg. chicken
seasoned cornflour for coating
2 tablespoons corn oil
1 onion, sliced
1 chicken stock cube
1 10-oz./283-g. can tomatoes, drained
1 green pepper, sliced
¼ teaspoon garlic salt
1 tablespoon paprika pepper
½ pint/3 dl. water

Joint the chicken and coat with seasoned cornflour. Heat the corn oil in a large saucepan and brown the chicken pieces on both sides. Remove from the pan. Add the onion to the pan and cook until tender. Add the chicken stock cube, tomatoes, pepper, garlic salt and paprika pepper. Stir in the water. Return the chicken to the pan, cover and simmer gently for 40–50 minutes or until the chicken is tender. Remove the chicken joints and boil the sauce in the pan rapidly until it is reduced by half.

To freeze: Cool rapidly, put into a suitable container, cover closely and seal before freezing.

To prepare for serving: Turn into a saucepan and reheat gently. Plain boil 12 oz. (350 g.) Patna rice and put it around the edge of the serving dish. Place the chicken joints in the centre. Add ¼ pint (1½ dl.) soured cream to sauce and reheat without boiling. Pour over the chicken. *Serves 6.*

Chicken chop suey

1 large head celery

3 large onions

salt and pepper to taste

1 tablespoon sugar

2 pints/generous litre chicken stock or
   2 stock cubes and water

1 can bean sprouts

1 can bamboo shoots

1 lb./450 g. cooked chicken, cut into fine strips

1½ oz./40 g. cornflour

3 tablespoons cold water

2 tablespoons soy sauce

Clean the celery and cut into thin strips. Peel and finely slice the onions. Simmer the celery, onion, salt, pepper and sugar in the stock and the liquid from the cans of bean sprouts and bamboo shoots for 20 minutes. Add the chicken, bean sprouts and finely chopped bamboo shoots to the vegetable mixture. Blend the cornflour with the cold water and stir into the mixture. Simmer gently for 10 minutes, stirring frequently. Add the soy sauce.

To freeze: Cool rapidly, pack in a large plastic container. Seal and freeze.

To prepare for serving: Thaw at room temperature allowing approximately 6 hours. Reheat gently in a covered saucepan. *Serves 6.*

Poulet sauté Grand Monarque

1 oz./25 g. butter

1 tablespoon cooking oil

6 chicken pieces

1 medium onion, peeled and finely chopped

6 button onions, peeled

6 button mushrooms, washed and trimmed

3 tablespoons brandy

¼ teaspoon tarragon

¼ pint/1½ dl. stock or water

½ teaspoon salt

¼ pint/1½ dl. dry white wine

Heat the butter and oil in a large frying pan. When very hot add the chicken pieces and fry to brown on all sides. Transfer the chicken to an ovenproof dish. Lower the heat and sauté the chopped onion until soft but not brown; add the button onions and cook until golden. Spoon the onions into the dish with the chicken. Fry the mushrooms gently for 2 minutes; set aside.

Pour the brandy over the chicken and set alight. When the flames have died down add the remaining ingredients, with the exception of the wine. Cover with a lid and cook in the centre of a moderate oven (350°F., 180°C., Gas Mark 4) for 45 minutes. Add the wine and continue to cook for a further 15 minutes.

To freeze: Add the mushrooms and cool rapidly. Put into a suitable container, cover closely and seal before freezing.

To prepare for serving: Turn into a saucepan and reheat gently. Correct the seasoning and thicken with 1 oz. (25 g.) beurre manié (see page 130). Lower the heat and simmer for 10–12 minutes. *Serves 6.*

Jugged hare

1 hare, skinned and jointed (reserve blood)

salt and pepper to taste

bouquet garni

2 stock cubes (optional)

seasoned flour

8 oz./225 g. streaky bacon, diced

3 carrots, peeled and sliced

1 large onion, stuck with 6 cloves

sprig thyme

pinch mace

salt and pepper

¼ pint/1½ dl. red wine

4 oz./100 g. redcurrant jelly

Use the carcass and head of the hare to make stock, seasoning well with salt, pepper and a bouquet garni, or use 2 beef stock cubes to 1¼ pints (7 dl.) boiling water.

Turn the joints in seasoned flour. Fry the bacon gently and set aside, retaining the dripping. Fry the joints of hare in the rendered fat in a saucepan or flameproof casserole to seal all surfaces. Add to the bacon with the carrots, onion, herbs and seasoning.

Pour in enough stock just to cover, put on the

lid, and simmer gently for 2–2½ hours according to the size and age of the hare. Ten minutes before it is ready, mix together the blood, red wine and redcurrant jelly and add to the hare, to thicken the sauce.

To freeze: Remove the joints and reduce the sauce slightly, if desired, by further simmering. Cool rapidly; either freeze in the casserole or put into some other suitable container. Cover closely and seal before freezing.

To prepare for serving: Turn into a saucepan, reheat carefully over gentle heat. *Serves 6.*

Casserole of pigeons

4 oz./100 g. pork

4 oz./100 g. veal

3 wood pigeons, cleaned

4 oz./100 g. streaky bacon, stretched

1 oz./25 g. seasoned flour

2 oz./50 g. butter

½ pint/3 dl. chicken stock (see page 103)

1 onion

1 carrot

6 parsley stalks

salt and pepper to taste

8 oz./225 g. mushrooms

Finely mince the pork and the veal together. Divide the mixture into three and put a portion in the body cavity of each bird. Cover the breasts with bacon and secure with string. Coat the birds with seasoned flour, melt 1 oz. (25 g.) of the butter in the casserole and brown the birds on all sides over a brisk heat. Add the stock, onion, carrot, parsley stalks and seasoning. Cover with a lid and bake in the centre of a moderate oven (325°F., 170°C., Gas Mark 3) for 2–2½ hours.

Remove the pigeons from the casserole dish; remove the bacon, cut into dice and set aside. Cut the pigeons in half down the centre. Melt the remaining butter in a small saucepan, add the mushrooms and cook for 1½ minutes.

To freeze: Put the pigeons into a small baking tin with the bacon, mushrooms and strained liquor. Cool rapidly. Cover and seal with foil before freezing.

To prepare for serving: Partially thaw at room temperature allowing approximately 2 hours. Reheat in a moderate oven (350°F., 180°C., Gas Mark 4). *Serves 6.*

Baked Alaska

1 8-inch/20-cm. sponge cake

1 tablespoon raspberry jam

1 family brick ice cream

Spread the base of the cake with the jam. Mash the ice cream with a fork and spread on top of the jam.

To freeze: Place on a baking sheet and freeze until firm, then wrap with foil.

To prepare for serving: Whisk 2 egg whites stiffly and gradually add 3 oz. (75 g.) castor sugar whilst still whisking. When the meringue is stiff and shiny, remove the base from the freezer, unwrap and spoon the meringue on top. Cook in the top of a pre-heated very hot oven (450°F., 230°C., Gas Mark 8) for 3–5 minutes or until brown. Serve immediately. *Serves 6.*

Rum sponge

3 eggs

6 oz./175 g. castor sugar

3 oz./75 g. plain flour

¼ pint/1½ dl. water

2 tablespoons rum

Put the eggs and half the sugar in a mixing bowl over a saucepan of hot water. Whisk until the mixture is thick enough to leave a trail. Sieve the flour on to the surface of the mixture and carefully fold it in.

Pour into a greased 8-inch (20-cm.) savarin tin and bake in the centre of a moderately hot oven (375°F., 190°C., Gas Mark 5) for 30–35 minutes or until firm to the touch and golden brown in colour. Turn out and cool on a wire tray.

Dissolve the remaining sugar in the water. Bring to the boil and continue to boil rapidly for 7 minutes; add the rum. Cool, then prick holes in the sponge with a skewer. Brush with the rum-flavoured syrup until it is all absorbed into the cake.

To freeze: Wrap tightly in double or heavy duty foil, polythene tissue paper or thick polythene. Seal and freeze.

To prepare for serving: Thaw, covered, at room temperature allowing approximately 8 hours. Unwrap and place on a serving dish. Top with peeled and quartered pears frozen in syrup. Use for decoration orange segments and glacé cherries. *Serves 6–8.*

Frozen mocha sponge pudding

1 4-oz./113-g. packet chocolate chips
4 tablespoons strong coffee
2 sponge cakes, baked in 7- by 11-inch/18- by 28-cm.
  Swiss roll tins
8 oz./225 g. unsalted butter
1 lb./450 g. icing sugar, sieved
2 egg yolks

Soften the chocolate in a basin, over a saucepan of hot water. When melted, gradually add the coffee. Trim the edges of both sponges and cut them across the width. Cream the butter until soft, then gradually work in the icing sugar and egg yolks. Beat the chocolate into the mixture.

Spread one piece of sponge with one-third of the chocolate filling, then build up the layers pressing each piece of sponge firmly in place on top of the filling. Finish with a sponge layer.

To freeze: Fit into a rectangular cake tin or container until frozen, then wrap in double or heavy duty foil and replace in the freezer.

To prepare for serving: Thaw wrapped at room temperature, for 4–5 hours. Sieve icing sugar over the top. *Serves 6.*

Lemon chiffon pie

1 lemon
2 eggs, separated
2 oz./50 g. castor sugar
2 teaspoons gelatine dissolved in
  2 tablespoons hot water
1 baked 9-inch/23-cm. flan case

Squeeze the lemon juice into a mixing bowl and add half the zest, finely grated. Add the egg yolks and sugar and beat, over a saucepan of hot water,

until light and fluffy. Remove from the heat and continue to whisk until cool. Whisk the egg whites stiffly. Whisk the gelatine into the egg yolk mixture and continue to whisk until it is on setting point. Quickly fold in the egg whites; turn the mixture into the flan case and chill until set. Partly freeze.

To freeze: Cover the filling with polythene tissue and wrap the pie in double or heavy duty foil to seal before returning to the freezer.

To prepare for serving: Thaw in wrapping, at room temperature, for 4 hours. Decorate with whipped double cream. *Serves 6.*

Lemon curd cloud

8 oz./225 g. lemon curd (see page 121)
¼ pint/1½ dl. single cream
2 egg whites, stiffly beaten

Stir the lemon curd into the cream until well blended, then fold in the egg whites.

To freeze: Pack in individual plastic dishes. Seal and freeze.

To prepare for serving: De-frost and serve with sponge fingers or langue du chat biscuits. *Serves 4.*

Note: The desserts can be served decorated with grated plain chocolate or a sprinkling of nutmeg.

Lemon soufflé

4 tablespoons cornflour
6 eggs, separated
1 lb. 2 oz./500 g. castor sugar
1½ pints/scant litre milk
6 lemons
¾ oz./20 g. gelatine, dissolved in
  9 tablespoons hot water

Blend the cornflour, egg yolks and sugar with a little of the cold milk. Put the remaining milk on to heat with the lemon rind, pared into thin strips. When the milk is almost boiling, strain off the lemon rind and pour the milk on to the blended cornflour, while stirring. Return the mixture to the pan, bring to the boil, and cook for 3 minutes, stirring all the time. Pour the mixture into a large bowl. Stir in the dissolved gelatine and juice from the lemons. Leave to cool. Whisk the egg whites until stiff. Fold into the cooled mixture.

To freeze: Pour into a suitable plastic container and leave in a cool place to set. Cover and freeze.

To prepare for serving: Remove the covering and thaw at room temperature allowing approximately 5 hours. Dip the base of the container into warm water and unmould on to a flat plate. Decorate with whipped cream. *Serves 8.*

Chocolate ice cream

4 oz./100 g. margarine

6 oz./175 g. castor sugar

6 oz./175 g. plain chocolate

4 eggs, separated

½ teaspoon vanilla essence

½ pint/3 dl. evaporated milk, scalded

Cream the margarine and sugar until light and fluffy. Melt the chocolate in a basin over hot water. Stir the chocolate into the egg yolk, add the vanilla essence and creamed mixture.

Whisk the evaporated milk and fold into the chocolate mixture. Pour into a plastic container, cover and freeze for 1 hour. Remove from the freezer and mash with a fork. Whisk the egg whites stiffly and fold into the mixture. Seal and return to the freezer. *Serves 6.*

Coffee ice cream charlotte

1 tablespoon apricot jam, sieved

19 Boudoir biscuits

For the coffee ice cream:

4 eggs, separated

4 oz./100 g. icing sugar, sieved

4 tablespoons bottled coffee essence

3 tablespoons rum

½ pint/3 dl. double cream, lightly whipped

Brush the sides of a 6-inch (15-cm.) cake tin with warmed apricot jam. Trim the biscuits to the depth of the tin then stand round the sides of the tin so they fit closely. Whisk the egg whites until very stiff, then gradually whisk in the icing sugar. Whisk the egg yolks, coffee essence and rum together then whisk gradually into the egg whites. Fold in the cream.

Chocolate ice cream

To freeze: Pour into the middle of the cake tin and freeze in the tin with the top tightly covered with foil.

To prepare for serving: Carefully loosen round the sides of the tin with a palette knife, then dip the bottom of the tin quickly in warm water. Turn out on to a serving plate. Decorate the top with whipped cream and toasted almonds. *Serves 6–8.*

Lemon sorbet

8 oz./225 g. granulated sugar

1 pint/6 dl. water

2 lemons

1 egg white

Dissolve the sugar in the water over a gentle heat. Pare the rind thinly from one of the lemons and add it to the sugar. Boil rapidly for 10 minutes, to form a light syrup. Strain into a mixing bowl, allow to cool, add the strained juice of both lemons and freeze for 2 hours. Remove the sorbet from the freezer and whisk to break down the ice particles. Whisk the egg white stiffly and fold into the sorbet.

To freeze: Spoon the mixture into a 1½-pint (1-litre) container. Cover closely with foil and re-freeze.

To prepare for serving: Serve in scoops while still frozen. *Serves 6.*

Strawberry cream torte

5 egg whites
pinch cream of tartar
8 oz./225 g. castor sugar
2 oz./50 g. flaked almonds, browned
For the filling:
1 tablespoon gelatine dissolved in
    2 tablespoons hot water
½ pint/3 dl. double cream
½ pint/3 dl. single cream
1 tablespoon castor sugar
3 tablespoons Curaçao
1 lb./450 g. fresh strawberries, sliced

Grease and flour two baking sheets. Trace two 6-inch (15-cm.) circles on each. Pre-heat the oven to 225°F., 110°C., Gas Mark ¼.

Whisk the egg whites and cream of tartar until fluffy, add a third of the sugar and continue to whisk until stiff. Add another third of the sugar and whisk until the mixture is shiny and heavy. Fold in the remaining sugar. Spread the mixture evenly within the traced circles and sprinkle with flaked almonds. Cook in the bottom of the oven for 2–3 hours or until crisp. Remove from the baking sheet and cool on a wire tray.

Dissolve the gelatine in the water and leave to set. Whisk the double and single cream until thick enough to leave a trail. Melt the gelatine in a small saucepan, without boiling; mix into the cream with the sugar and the Curaçao. Stir until on the point of setting. Sandwich the meringues together, with cream and strawberries between each layer, reserving some cream to coat the sides and top of the torte.

To freeze: Place on a baking sheet and partially freeze in order to set the cream before over-wrapping with sheet polythene.

To prepare for serving: Unwrap and serve frozen. *Serves 6–8.*

Frozen Christmas pudding

For the one-stage chocolate cake:
2 oz./50 g. luxury margarine
2 oz./50 g. castor sugar
2 tablespoons rum (optional)
2 oz./50 g. self-raising flour sifted with
    ½ teaspoon baking powder
1 heaped tablespoon cocoa blended with
    1 tablespoon hot water
1 large egg
For the ice cream:
2 oz./50 g. castor sugar
¾ oz./20 g. gelatine dissolved in
    4 tablespoons water
1 16-oz./454-g. can evaporated milk, chilled
¼ teaspoon almond essence
2 oz./50 g. glacé cherries, halved
little red colouring
2 oz./50 g. sultanas and currants, mixed
2 teaspoons lemon juice
grated rind of 1 lemon

To make the cake, put all the ingredients into a bowl and beat with a wooden spoon for about 3 minutes. Put into a greased and lined 7-inch (18-cm.) sandwich tin and bake in the centre of a very moderate oven (325°F., 170°C., Gas Mark 3) for 25–30 minutes. Remove from the tin and cool.

While the cake is cooling, prepare the ice cream. Add the sugar and dissolved gelatine to the chilled milk, whisk and put in the refrigerator until just beginning to set. Whisk again and divide in half. Add the almond essence and cherries to one half, then colour pink. Add the remaining ingredients to other half. Pour the pink mixture into a 3-pint (1¾-litre) pudding basin, cover with the cake. Put the other ice cream on top.

To freeze: Cover the top of the basin with foil and seal.

To prepare for serving: Put the basin into hot water for about 1 minute, remove the foil lid, invert and turn on to a serving plate. Decorate with holly and serve with whipped cream, topped with cinnamon.

For smaller parties, cook the cake mixture in two 5-inch (13-cm.) tins for 20 minutes. Divide the ice cream mixture between two 1½-pint (1-litre) bowls. Serve one pudding and store the other in the freezer.

Apple and chocolate flan

5 oz./150 g. butter
5 oz./150 g. castor sugar
2 large eggs, well beaten
4 oz./100 g. plain flour
1 oz./25 g. cocoa powder
1 teaspoon baking powder
pinch salt
¼ pint/1½ dl. milk

Cream together the butter and sugar until light and fluffy. Gradually beat in the eggs. Sieve together the flour, cocoa powder, baking powder and salt and fold into the sponge mixture. Stir in the milk and turn into a greased 8½-inch (21-cm.) flan tin. Bake in a moderately hot oven (375°F., 190°C., Gas Mark 5) for 45 minutes or until the sponge is firm to the touch. Cool on a wire tray.

To freeze: Wrap closely in foil and freeze.

To prepare for serving: Thaw, covered, at room temperature allowing approximately 3 hours. Poach partially thawed apple slices in a sugar syrup until tender. Drain and cool. Just before serving, fill the centre of the flan with scoops of ice cream, top with poached apple slices and decorate with grated chocolate. *Serves 6.*

Strawberry soufflé

12 oz./350 g. fresh strawberries
castor sugar
½ oz./15 g. gelatine dissolved in ¼ pint/1½ dl. cold water
4 eggs, separated
red colouring
¼ pint/1½ dl. double cream, lightly whipped

Tie a greased, double thickness band of greaseproof paper round a 6-inch (15-cm.) soufflé dish. Hull and wash the strawberries. Sieve or put into a blender and sweeten the purée to taste. Put the gelatine and water in a basin over a pan of gently simmering water until the gelatine has dissolved.

Whisk the egg yolks with 3 oz. (75 g.) sugar over a pan of gently simmering water until thick and creamy. Remove from the heat and whisk until cool. Stir in the gelatine and strawberry purée and colour with a little red colouring to give a good pink colour. When the mixture is thick but not set, fold in the whipped cream. Stiffly whisk the egg whites and fold into the mixture. Turn into the prepared soufflé dish. Put in the freezer and partially freeze.

To freeze: Remove the greaseproof collar, using a knife dipped in hot water, if necessary. Cover with polythene tissue paper, seal and replace in the freezer.

To prepare for serving: Allow to thaw in the refrigerator for 24 hours. Decorate with whipped cream and strawberries. *Serves 6.*

Apple and blackcurrant mould

½ pint/3 dl. double cream
½ pint/3 dl. frozen apple purée, thawed
1 egg, separated
1 oz./25 g. gelatine
¼ pint plus 4 tablespoons/2 dl. water
1 lb./450 g. blackcurrants
sugar to taste
1 dessertspoon cornflour

Whip the cream until thick enough to leave a trail then fold into the apple purée, with the stiffly beaten egg white. Dissolve ½ oz. (15 g.) of the gelatine in 2 tablespoons water, heat gently to melt, pour into the apple mixture and stir until on the point of setting. Pour into a large mould and place a small mould of the same shape in the centre. Place in the refrigerator and allow to set before removing the smaller mould.

Meanwhile, simmer the blackcurrants in ¼ pint (1½ dl.) water with sufficient sugar to taste. Blend the cornflour with a little of the juice, add to the blackcurrants, bring to the boil and simmer for 1 minute, stirring constantly. Cool and beat in the egg yolk. Allow the mixture to become completely cold.

Dissolve the remaining gelatine in the rest of the water and heat gently to melt. Add to the blackcurrant mixture and stir until on the point of setting. Pour into the mould and return to the refrigerator to set.

To freeze: Seal tightly with double or heavy duty foil and freeze.

To prepare for serving: Thaw overnight in the refrigerator. Cut into slices and serve with Boudoir biscuits. *Serves 6–8.*

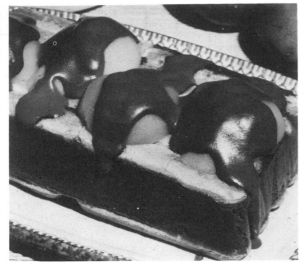

Chocolate cake with pears and
chocolate sauce

Chocolate cake with pears and chocolate sauce

3 oz./75 g. Savoy fingers or dry sponge cake

1 oz./25 g. plain chocolate

4 oz./100 g. castor sugar

2 eggs, separated

2 tablespoons milk

½ oz./15 g. gelatine

1 oz./25 g. butter

½ teaspoon vanilla essence or 1 teaspoon rum

2 dessert pears, peeled and cored

Use either an ice tray about 7–8 inches (18–20 cm.) long or a 1-lb. (450-g.) loaf tin (the shallow aluminium type). Cut a piece of non-stick lining paper to line the bottom and sides of the tin and give sufficient length to fold over the finished mixture.

Split the Savoy fingers or cut the sponge cake into thin slices. Put a layer in the bottom of the prepared tray, with the rounded sides of the Savoy fingers next to the paper. Keep the remaining pieces for the top.

Melt the chocolate in a basin over hot water or in a saucepan over a gentle heat. Add half the sugar, the egg yolks and the gelatine dissolved in the milk and continue cooking over a gentle heat stirring all the time, until the mixture thickens. Cool, stirring occasionally.

Cream together the butter and remaining sugar until soft and light. Add to the chocolate mixture and combine thoroughly. Add the vanilla essence or rum. Beat the egg whites until stiff and fold them into the mixture. Pour it into the prepared tray and put another layer of fingers or cake on top. Fold over the spare paper to cover the top. Peel the pears, halve and remove the cores. Poach the pears in heavy sugar syrup until tender.

To freeze: Cover the cake with double or heavy duty foil. Seal and freeze. Cool the pears quickly and put into a plastic container, leaving a ½-inch (1-cm.) headspace; cover with crumpled foil to keep the pears submerged. Cover and freeze.

To serve: Thaw in the refrigerator for 12 hours and turn out the cake while still slightly chilled. Top with the pears and pour over Dark chocolate sauce (see page 96). *Serves 6.*

Banana cream

6 bananas

¼ pint/1½ dl. double cream, lightly whipped

¼ pint/1½ dl. thick custard (1 tablespoon custard powder to ¼ pint/1½ dl. milk)

¼ teaspoon vanilla essence

juice of ½ lemon

2 oz./50 g. castor sugar

½ oz./15 g. gelatine dissolved in ¼ pint/1½ dl. hot water

Mash the bananas to a purée with a fork. Blend the banana, whipped cream and cold custard. Add the vanilla essence, lemon juice and gelatine. Whisk until on the point of setting.

To freeze: Pour into a plastic or foil container. Cover closely and freeze.

To prepare for serving: Thaw at room temperature for approximately 4 hours. *Serves 6.*

Index

142